Fluid Power
Educational
Series

Hydrostatic Transmissions (HSTs)

(In the SI Units)

Joji Parambath

Hydrostatic Transmissions (HSTs)
(In the SI Units)

Copyright © 2026 Joji Parambath

All rights reserved

ISBN: 9798859104079

https://jojibooks.com

First Edition – 2020
Revised Edition – 2021
Second Edition – 2023
Revised Edition 2026

Disclaimer of Liability

The contents of this book have been checked for accuracy. We cannot guarantee full agreement, as deviations cannot be entirely precluded. Only qualified personnel should be permitted to install and service hydraulic equipment. Qualified persons are those authorised to commission, ground, and tag circuits, equipment, and systems in accordance with established safety practices and standards.

Table of Contents

PREFACE

Hydrostatic transmissions (HSTs) occupy a distinct place in fluid power technology. They are widely used in agricultural tractors, on- and off-highway mobile equipment, and various self-propelled equipment for power transfer and control. An HST provides an infinitely variable speed between zero and maximum in both forward and reverse modes of operation without changing the prime mover's speed. Moreover, manufacturers are bringing out smaller, lighter HSTs with advanced electronic controls and improved performance. These factors make HSTs a cost-effective choice for many industrial and mobile applications.

The second edition of this introductory book presents the principles of hydrostatic transmissions. The basic concepts of typical open-circuit and closed-circuit HSTs are described in a simple-to-understand manner. The configurations, types, specifications, and applications of HSTs are also given. Appendix 2 contains case studies on an open-circuit hydraulic concrete pump system, a hydraulic steering system, and typical displacement controls of a bidirectional variable-displacement axial-piston pump. Appendix 4 describes basic displacement-control methods for variable-displacement axial-piston pumps. The book uses the SI system of units.

This book provides educational information designed to equip hydraulic professionals to handle complex hydraulic circuits. Every HST circuit for real-world applications is unique and must be developed and tested to ensure the required performance and life.

The author presents many fluid power topics in other textbooks under the fluid power educational series. A list of all the textbooks is given at the end of the book. Also, please see the details at https://jojibooks.com.

Enjoy reading the book.
Your feedback is most welcome.

JOJI Parambath

Chapter 1 | Introduction to Hydrostatic Transmissions (HST)

In modern industrial and mobile systems, power must be transmitted from prime movers to industrial machines and mobile equipment in a controlled manner. Power transmission can generally be achieved mechanically, electrically, or hydrostatically.

The mechanical arrangement for power transmission uses gears, clutches, and belt drives. In mechanical power transmission, the prime mover's spatial position relative to the load is fixed, making stepless control of the drive difficult.

An electric motor-powered drive is called an electric drive. It can provide a broad range of torque, speed, and power. However, a major drawback of the electric drive is that it cannot be used in situations where power is not readily available, such as in mobile systems.

In a hydrostatic transmission (HST), a hydraulic pump drives a hydraulic motor. The hydraulic motor's speed, torque, or power can be regulated hydrostatically. The overwhelming majority of today's HSTs use variable-displacement pumps and fixed-displacement motors. Remember that IC engines usually drive HSTs in mobile hydraulic systems.

Depending on its design, an HST can drive a load from the maximum speed in one direction to the maximum speed in the opposite direction, with a fine variation of the speed between the two maximums – all with the prime mover of the pump running at a constant speed. Many systems exploit this advantage. However, a significant drawback of HSTs is their high cost.

In both industrial and mobile applications, HSTs come in two types:

- Open-circuit (open-loop) drives

- Closed-circuit (closed-loop) drives

This book aims to explain hydrostatic transmission concepts in an easy-to-understand way.

Chapter 2 | Open-circuit HSTs

Most general-purpose hydraulic systems, especially those involving linear actuators, are open-circuit designs. An open-circuit hydrostatic transmission (HST), as shown in the schematic diagram in Figure 2.1(a), consists of a prime-mover-driven pump, a load-coupled hydraulic motor, a reservoir, and all the required controls in one package. All HST elements are interconnected via metal tubing or hose assemblies.

The driven pump draws the fluid directly from the reservoir, generates the necessary flow, and drives the motor. The fluid is then returned to the reservoir. Next, the motor's rotation direction can be reversed using a directional control valve, as shown in the circuit diagram in Figure 2.1(b).

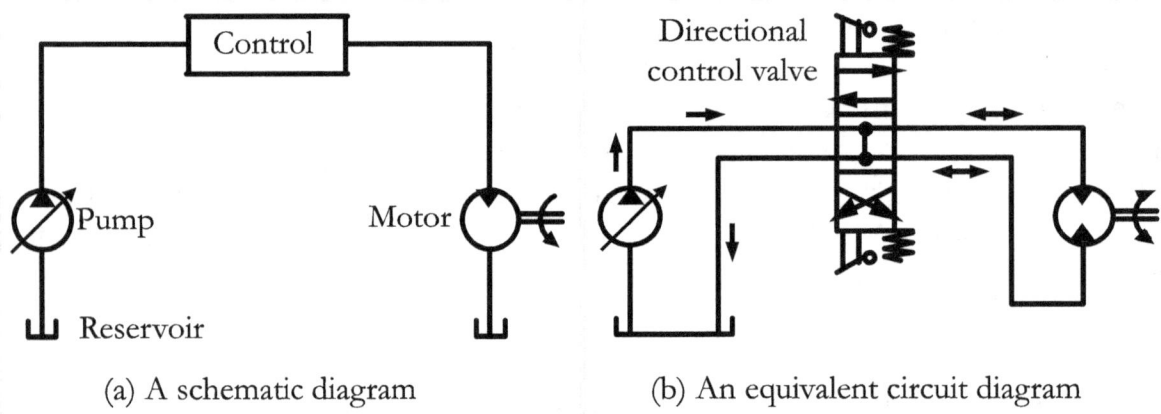

(a) A schematic diagram   (b) An equivalent circuit diagram

Figure 2.1 | An Open-circuit HST

The open-circuit arrangement can provide infinitely variable speed, torque, or power, ranging from zero to maximum values in both the motor's forward and reverse directions of motion.

Open-circuit systems are cheaper and easier to maintain than closed-circuit ones. However, the reservoir must be sufficiently large to cool the fluid effectively.

Open-circuit systems are used in industrial machinery, where there is no space or weight constraints. An open-circuit hydraulic concrete pump system is explained in the following sections.

Open-circuit Hydraulic Concrete Pump System

A concrete pump system continuously transfers freshly-mixed concrete to desired positions in a construction site using the power and flexibility of open-circuit hydraulics and a valve system. It typically consists of unidirectional hydraulic pumps, a hopper, a twin-cylinder system with two drive cylinders and two material cylinders, a rock valve, a slewing cylinder, control valves, an accumulator, and an agitator. The following sections briefly explain these components.

Bent-axis axial-piston variable-displacement piston pumps are usually employed to supply the required fluid to the system. The displacement and, hence, the flow rate of the hydraulic pump can be varied by using a displacement control system. The concrete mixture is transferred to a hopper, where it is continuously churned to prevent solidification.

Pump Displacement Controls

In a hydraulic system with a variable-displacement axial piston pump, it is necessary to control pump displacement to vary the flow rate and limit system pressure to the design value. Different methods can change the displacement of an axial piston pump with a swash plate design. Some important methods for pump displacement control, such as the pressure controller (limiter) and the pressure controller with load sensing, are briefly presented at the end of the chapter.

Drive Cylinders and Material Cylinders

A typical concrete pump system consists of two identical drive cylinders, each with an attached material cylinder. Each material cylinder has a rubber ram (piston) connected to the piston rod of the corresponding drive cylinder. A material cylinder can draw and then push the concrete mixture when driven by the drive cylinder. Figure 2.2 shows a typical configuration of drive cylinders and material cylinders.

Typically, the piston side ports of the drive cylinders are connected. The fluid enters or exits through the rod side ports. A drive cylinder retracts when fluid enters its rod side port. The fluid exiting from this cylinder's piston side port enters the second cylinder's piston side port, and the driven cylinder extends. The fluid can exit the rod-side port of the second cylinder into the tank. It can be observed that when one drive cylinder retracts, the other extends simultaneously.

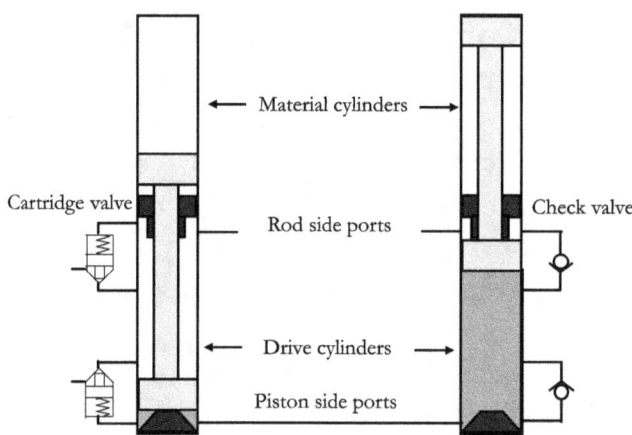

Figure 2.2 | A typical configuration of drive cylinders and material cylinders

However, one cylinder may experience a short stroke due to leakage, while the other completes its stroke. This problem can be eliminated by installing sensors and check valves on the cylinder.

When a drive cylinder retracts, concrete is drawn from the hopper into the associated material cylinder while the other drive cylinder extends. The extending cylinder pushes concrete into the associated material cylinder, which then delivers it to the delivery lines through the opening of the rock valve. This arrangement alternately pumps the concrete. Sensors continuously trigger the material-pumping action via a controller.

Directional Control Valves: They control the direction of motion of the drive cylinders and shift the position of the rock valve to the fully loaded material cylinder.

Relief Valve: The valve protects the hydraulic circuit by relieving pressure before it reaches the set value.

Other Components: Apart from these components, the system may include the following components: accumulators and safety blocks, a variable-displacement pump with load sensing control for placing the boom, cylinders on the boom section, outriggers, boom-slewing gear, water pump, etc. Filter/strainer, heat exchanger, pressure switches, check valves, unloading valve, throttle check valves, pump compensator valve, brake valves, reservoir, etc., can also be included.

Swing (Slewing) cylinder: Since there are two material cylinders and one delivery line, the rock valve (S-tube) must be shifted to connect the delivery line to the loaded material cylinder from the emptied material cylinder. A through-piston-rod swing cylinder with grooves in the piston rod for the switching function can shift the rock valve between the two material cylinders. Figure 2.3 shows two positions of a swing cylinder with a rock valve.

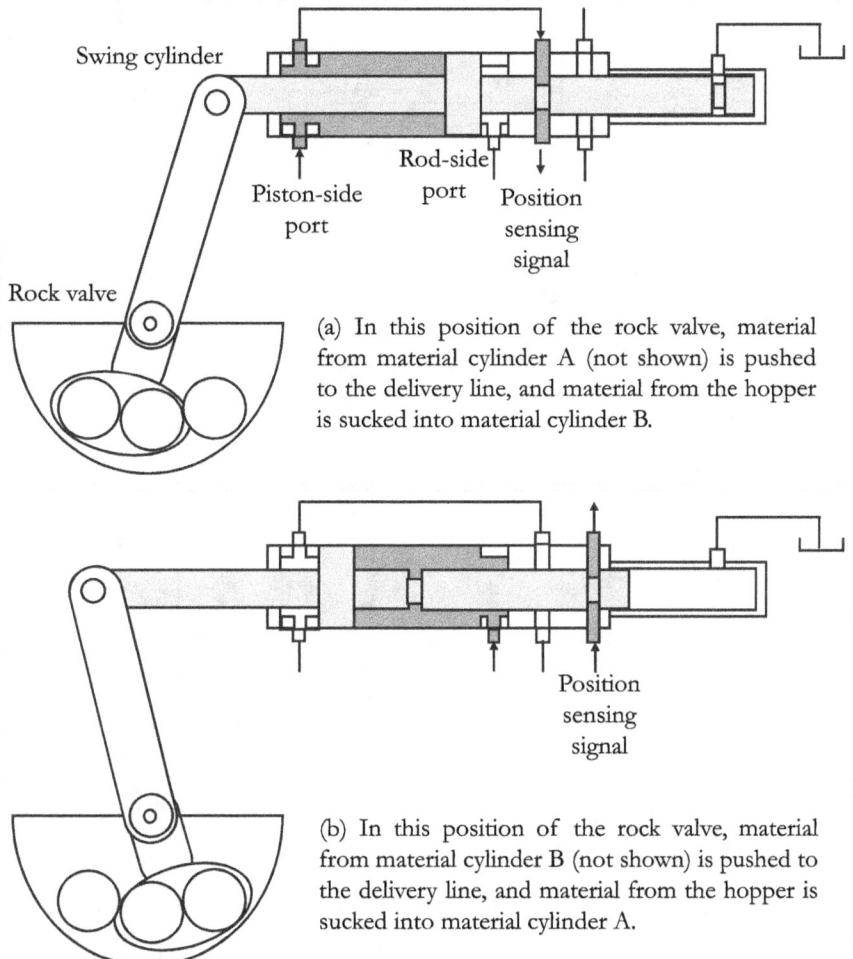

(a) In this position of the rock valve, material from material cylinder A (not shown) is pushed to the delivery line, and material from the hopper is sucked into material cylinder B.

(b) In this position of the rock valve, material from material cylinder B (not shown) is pushed to the delivery line, and material from the hopper is sucked into material cylinder A.

Figure 2.3 | Multiple positions of a Swing cylinder and a rock valve

The circuit diagram and brief explanation of an open-circuit hydrostatic concrete pump system are presented in the following section. A detailed explanation of the circuit, including multiple positions, is provided under Case Study #1 in Appendix 2.

An Open-Circuit Hydrostatic Concrete Pump System

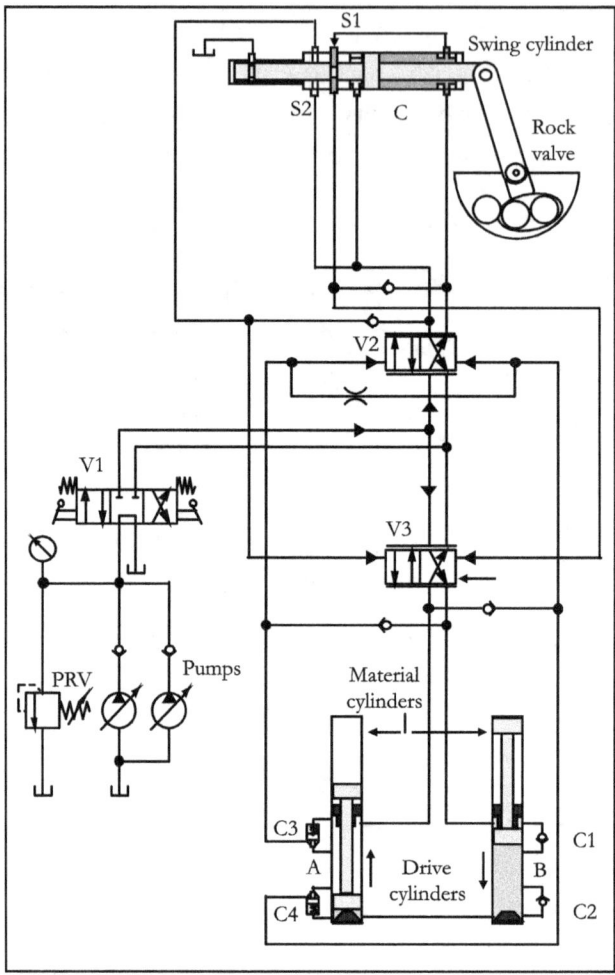

Figure 2.4 | An open-circuit hydrostatic concrete pump system.

Figure 2.4 shows a simplified open-circuit HST system for a concrete pump, consisting of unidirectional hydraulic pumps, a twin-cylinder system with two drive cylinders and two material cylinders, a rock valve, a slewing cylinder, and control valves. The circuit is presented here for an initial familiarization with the system.

Please note that Appendix 2 presents a case study of an open-circuit hydrostatic concrete pump system with multiple circuit positions, offering a more in-depth explanation of the circuit.

Pressure Controller for a Variable-displacement Axial Piston Pump

We know that a high-pressure relief valve can control pressure in a hydraulic system, but it dissipates significant energy as heat.

To prevent this, a pressure controller (also known as a pressure compensator) with a servo system is often used in systems with a variable-displacement pump. This helps regulate pump pressure and displacement without generating excessive heat in the system. Figure 2.5 presents a cross-sectional view and a schematic of a pressure controller.

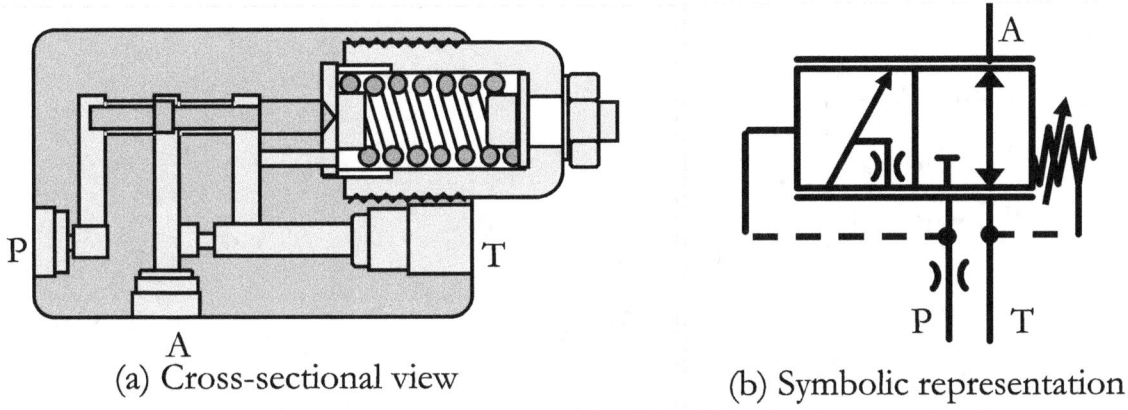

(a) Cross-sectional view   (b) Symbolic representation

Figure 2.5 | Pressure controller for variable-displacement axial piston pumps

A pressure controller with a servo system, associated with a variable-displacement pump, can effectively limit the maximum pressure at the pump outlet. It adjusts the pump's swash plate angle to match the system flow to the required pressure, thereby maintaining a constant pressure regardless of demand changes up to the pump's maximum capacity.

When system pressure increases, the pressure controller actuates the control piston, lowering the swash plate angle and reducing the flow rate.

Once the pressure reaches its maximum, the pressure controller automatically moves the swash plate to its lowest angle, reduces the flow rate to a minimum, and maintains the system pressure at the desired maximum.

The pressure setting is usually adjustable via a screw mechanism.

Concepts of Pump Displacement Control with Load Sensing

A load sensing system delivers only the flow at the required pressure to a system using a load sensing type pump, a load sensing compensator block, and a load sensing type directional control valve.

Load Sensing Type Variable-displacement Unidirectional Pump

The load sensing type variable-displacement unidirectional axial piston pump, as shown in Figure 2.6, can deliver variable flow and pressure as the hydraulic system dictates.

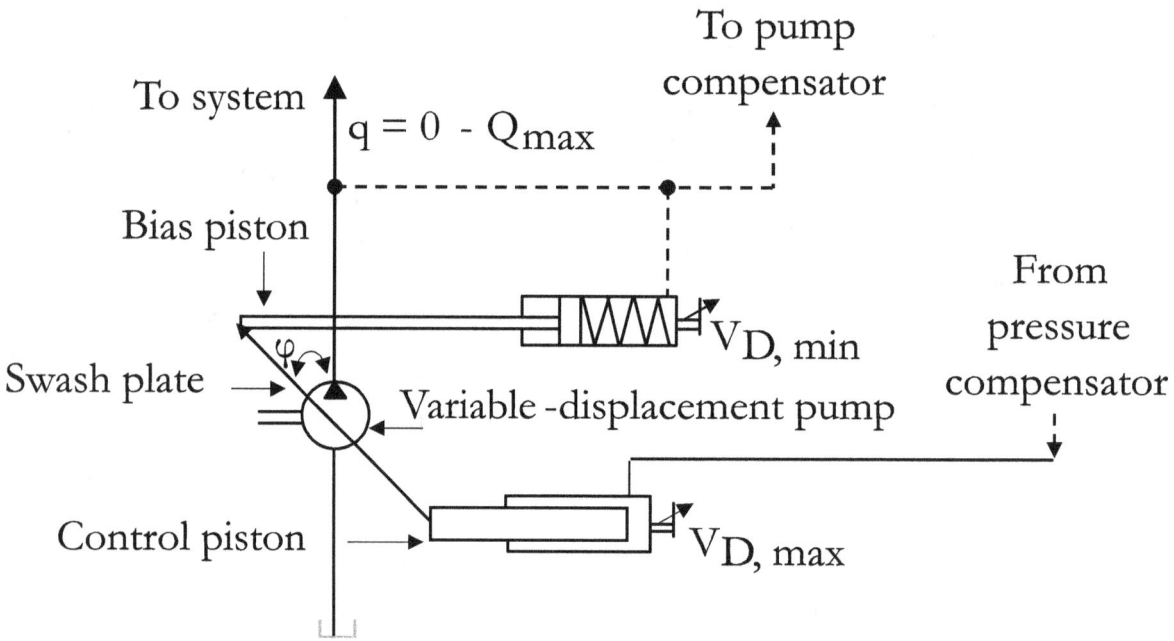

Figure 2.6 | A load sensing type variable-displacement unidirectional pump

The pump consists of a swash plate whose angle (φ) can be varied by bias and control pistons. A flow from the compensator into the control piston decreases the swash plate angle, decreasing the pump's displacement and flow output.

Different methods can be used to vary the displacement of an axial piston pump. That is, the displacement angle of a variable-displacement pump can be controlled mechanically by a lever attached to the swash plate, or servo-controlled via a displacement control valve (pressure-limiting compensator). Appendix 4 presents several displacement control methods.

Pump compensator block

A compensator senses the pressure and flow requirements of a load sensing system and makes the piston pump react to both variables properly. It contains a high-pressure compensator spool that works against, say, a 200-bar spring and a pressure-flow compensator spool that works against, say, a 20-bar spring, as shown in Figure 2.7(a).

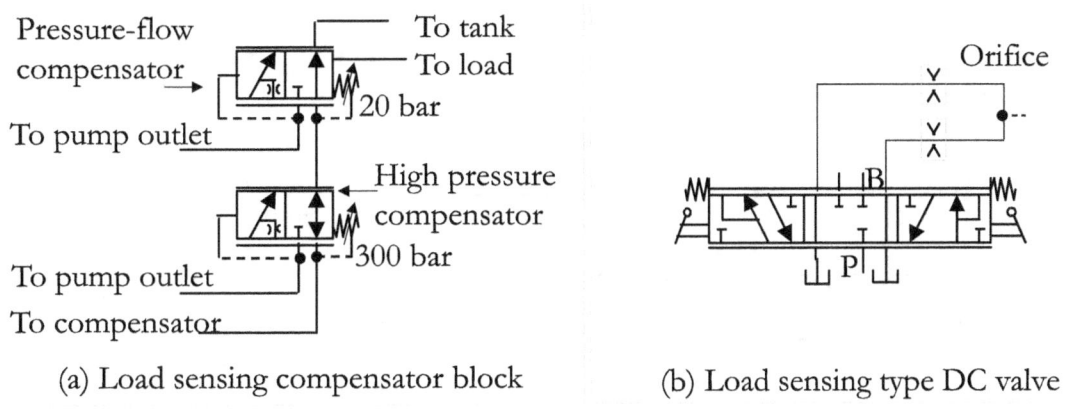

(a) Load sensing compensator block   (b) Load sensing type DC valve

Figure 2.7 | Load sensing components

Directional Control Valve, Load sensing type

The directional control valve is a closed-center position valve with a proportional characteristic. The symbolic representation of a load sensing type directional control valve is given in Figure 2.7(b). The valve also incorporates check valves (not shown) for proper operation of the load sensing system.

A Load Sensing Hydraulic System for an Open-circuit Hydraulic System

A load-sensing system for controlling a double-acting hydraulic cylinder, using a load-sensing pump, a compensator block, and a load-sensing directional control valve, is shown in Figure 2.8. An orifice must be installed between the pump and the cylinder to ensure proper system operation. The orifice is usually built into a directional control valve.

The load sensing system for controlling the double-acting cylinder hydraulic consists of a load sensing variable-displacement pump, a compensator block, and a load-sensing closed-center DC valve. The 20-bar spring of the compensator forces the pressure-flow compensator spool toward the left when there is no pressure. This normal spool position provides the fluid with a direct passage from the control piston to the reservoir. Because no pressure acts on the control piston, the swash plate moves to its maximum angle. The pump is ready to produce the maximum flow in the normal position.

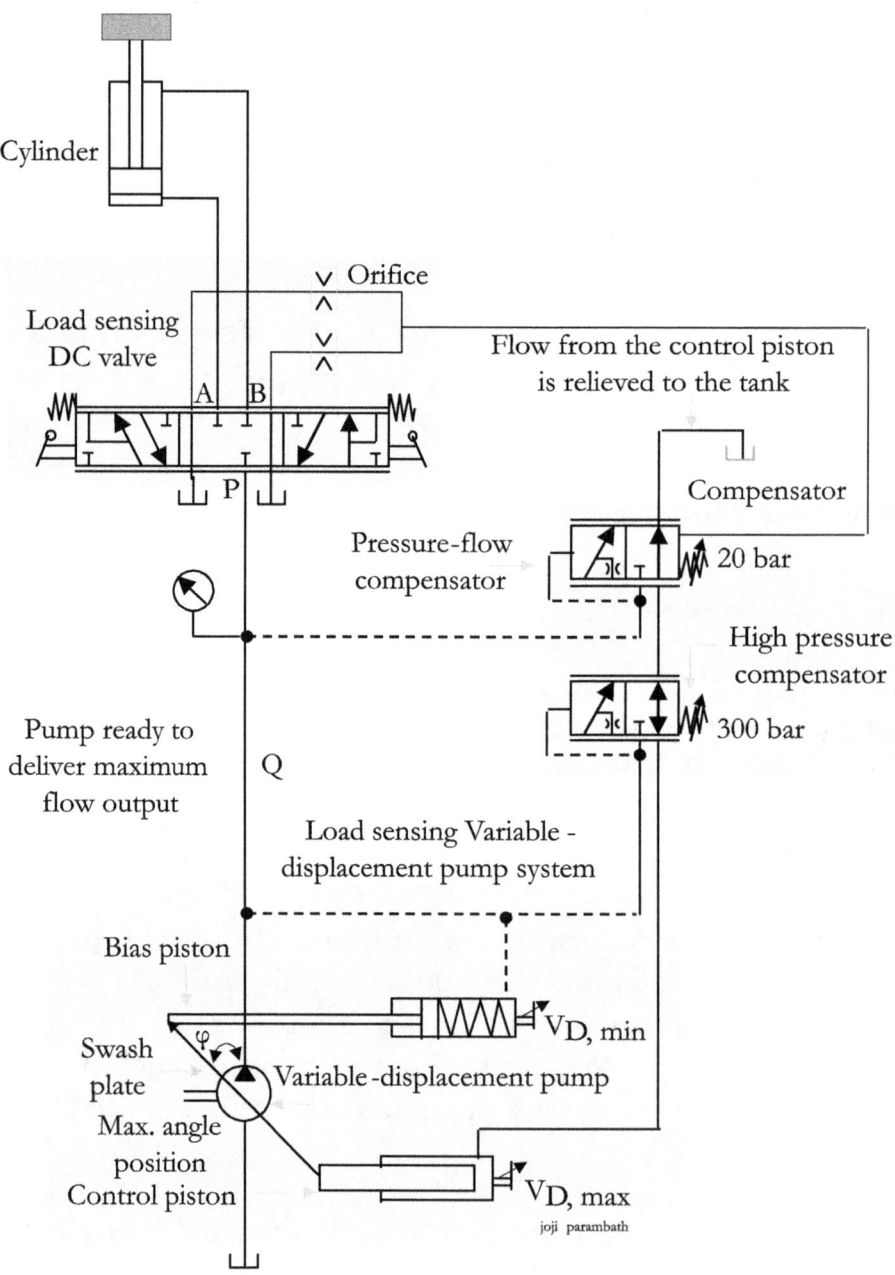

Figure 2.8 | The circuit of a load-sensing system in its initial position

The working of this circuit can be explained in three phases: (1) Low-pressure standby mode, (2) Load sensing mode, and (3) High-pressure standby mode.

Low-pressure Standby Mode: The pump flow acts on the left-hand side of the pressure-flow compensator spool and the high-pressure compensator spool when the pump is switched on. When the pressure reaches slightly above 20 bar, the pressure-flow compensator spool moves to the right against the low-pressure spring, directing flow to the control piston. This compensator flow causes the control piston to extend, thereby swinging the swash plate to its minimum angle and delivering minimum flow at low pressure. In standby mode, the pump provides only enough flow at low pressure to compensate for internal system leakage.

Load Sensing Mode: When the DC valve is shifted to the left envelope position, the flow is directed to the cylinder and develops pressure depending upon the load. The load pressure and the force of the 20-bar spring act on the RHS of the pressure-flow compensator spool. The pump outlet pressure acts on the LHS of the spool. The spool moves proportionally and remains in position based on the load pressure, directing a specific flow to the control piston. The control piston controls the position and, hence, the angle of the swash plate. As the swash plate angle decreases, the pump's flow output decreases; vice versa.

High-pressure Standby Mode: When the cylinder reaches its end-of-stroke position, the flow to the cylinder tends to stop, and the system pressure increases. Finally, when the flow stops, the pressures on both sides of the pressure-flow compensator spool equalize, and the pressure-flow compensator valve moves to its right-hand-side envelope position. When the pump outlet pressure reaches 300 bar, the high-pressure compensator valve moves to the left-hand envelope position, directing fluid to the control piston. The piston moves the swash plate to near-zero angles, and the pump ceases to produce flow. This is referred to as the high-pressure standby mode.

More Information: Additional information on displacement control is provided in Appendix 3. Further, the author elaborates on load sensing systems in a book entitled 'Load Sensing Systems (In the SI Units)' and 'Hydraulic Circuits - Identification of Components and Analysis'.

Chapter 3 | Closed-circuit HSTs

A closed-circuit hydraulic transmission (HST) system, with a pump and a hydraulic motor, directs the pump flow to the hydraulic motor inlet. The fluid discharged from the motor outlet flows directly back to the pump inlet, forming a power-transmission loop with high- and low-pressure sides. The system typically includes a pump, a motor, a charge pump, check valves, a shuttle valve, pressure relief valves, accumulators, and filters. Let's look at the details of closed-circuit hydrostatic transmission systems.

Basic Circuit of the Closed-circuit HST

The basic circuit with a pump and hydraulic motor is shown in Figure 3.1. Generally, a variable-displacement piston pump drives a fixed-displacement piston motor hydraulically. Case drain lines are provided on the pump and motor to relieve leakage.

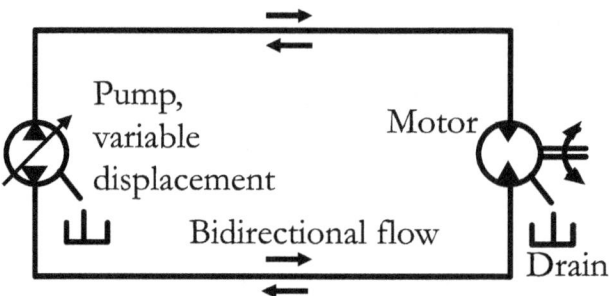

Figure 3.1 | The basic closed-circuit HST

Pump Unit

A variable-displacement bi-directional inline axial-piston pump of the swashplate construction or bent-axis design, driven at a constant speed, can serve as a power source. The swashplate angle, or the bent axis, can be infinitely varied using direct or servo-control displacement methods, yielding displacements from zero to maximum on both sides. The pump output flow and the associated hydraulic motor speed remain proportional to the pump displacement. Therefore, stepless adjustment of the swashplate angle or the bent axis can realise infinitely variable flow and hydraulic motor speeds. The direct-displacement control method uses a lever attached to the swashplate, while the servo control method utilises a servo piston attached to the swashplate and a feedback mechanism.

The maximum pressure in the system can be controlled by adjusting the pump's pressure compensator. The bidirectional pump allows the hydraulic motor to be driven in either direction.

Hydraulic Motor Unit

A fixed-displacement axial-piston motor of swash-plate construction with preset displacement can be used as the driven unit. The motor's speed is proportional to the flow rate of the input fluid. The direction of rotation of the motor's shaft depends on the direction of the fluid entry into the motor. The motor's output torque is directly proportional to the pressure differential across the motor.

Basic Features of Closed-circuit HSTs

The motor speed in a closed-circuit HST can be varied by adjusting the pump displacement and the flow rate. The motor's power output can be adjusted by changing the flow rate at a given pressure setting. The direction of rotation of the motor can be reversed by reversing the angle of the pump's swashplate.

The HST system will respond to increased load and torque by increasing the loop pressure. The motor's maximum output torque depends on the maximum pressure setting in the pump displacement control system.

Closed-circuit HSTs are available as fully integrated units, with all control elements enclosed in a compact housing. They are suited to applications requiring compact, lightweight, and power-packed hydraulic systems, such as mobile equipment and wind turbines.

Shortcomings of Closed-circuit HSTs

As the pump and motor in an HST leak internally due to excessive wear, fluid is lost from the transmission loop. Therefore, the pump and motor units must have case drain lines to return fluid to the reservoir and prevent leakage caused by pressure buildup behind the piston.

The basic closed-circuit HST system also suffers from excessive heat generation in the fluid. Therefore, high-quality piston pumps and motors are typically preferred in HSTs, as they minimize internal leakage and frictional losses.

A Closed-circuit HST with a Charge Pump

The transmission loop of a closed-circuit HST is affected by fluid loss from internal leaks in the pump and motor. As a preliminary step, a charge-pressure circuit can be added to a basic unidirectional hydrostatic drive, as shown in Figure 3.2(a), to compensate for leakage flows from the pump and motor case drain ports. The charge pump circuit consists of a properly sized charge pump, reservoir, check valve, pressure relief valve, and cooler.

This leakage-compensation concept can be extended to bidirectional operation of the main pump. An additional check valve is inserted into the charge pump circuit, as shown in Figure 3.2(b), to direct the fluid flow to the low-pressure side of the main transmission loop. The function of the charge pump is further elaborated in the following section.

Charge Pump

As indicated in the previous section, the basic charge pressure circuit consists of a low-pressure charge pump, a small fluid-filled reservoir, a relief valve, and two check valves.

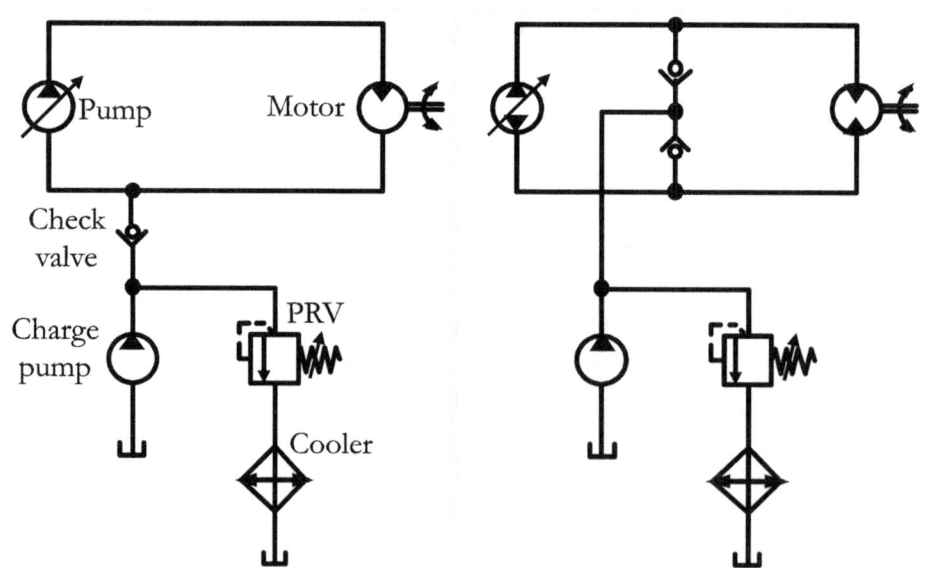

(a) For unidirectional pump operation   (b) For bidirectional pump operation

Figure 3.2 | HST circuits with charge pumps

A charge pump is essential in a closed-circuit HST. The smaller pump in the HST system draws fluid from the reservoir and delivers it to the loop. It is usually a fixed-displacement gear/gerotor pump mounted on the same main pump shaft.

The charge pump continuously adds fluid to the low-pressure side of the transmission loop via one of the open check valves, replenishing the fluid lost. The charge pump must provide sufficient makeup fluid to compensate for leakage, and a small percentage of the fluid must be purged from the main transmission loop through a flushing valve for flushing and cooling.

Furthermore, the charge pump must keep the low-pressure side of the loop filled with fluid to maintain the correct pressure. Consequently, it pressurises the main pump inlet port and prevents cavitation.

The charge pump must be appropriately sized. A rule of thumb is that the charge pump flow rate must be at least 20% of the main pump flow rate. That means the overall volumetric efficiency of the transmission loop can drop to 80% before cavitation occurs.

Reservoir

A reservoir of adequate size must be provided in the charge pump system to supply the charge pump with sufficient fluid and to cool it.

Typically, the reservoir size (in litres) for closed-circuit applications should be between 0.5 and 1.5 times the maximum flow rate (in lpm) delivered by the charge pump, and the fluid volume should be approximately 80% of the total reservoir capacity.

Charge Pressure Relief Valve (Charge PRV)

The charge pressure relief valve (PRV) is a low-pressure valve that maintains the pressure on the low-pressure side of the loop at the proper level. Once the loop is charged to the PRV's pressure setting, the flow from the charge pump passes over the PRV on its way back to the reservoir. Typically, the PRV is set to 10-35 bar, or in accordance with the manufacturer's specifications.

Note that using an excessively large charge pump can increase energy loss and the heat load on the transmission.

Check Valves

In applications where the load could exceed the motor's inertia rating, fluid loss from the transmission loop would cause cavitation. A check valve can prevent fluid loss from the loop. At the same time, the check valve permits the flow from the charge pump to the low-pressure side of the loop. The positive pressure at the pump's inlet eliminates cavitation. Two check valves are used to enable bidirectional operation of the loop.

Heat Exchanger / Cooler

A hydrostatic transmission system becomes hot when operated continuously under heavy load for extended periods. Therefore, hydrostatic transmissions generally require a properly sized heat exchanger (cooler) to remove excessive heat from the pump, motor, and transmission loop. A hydrostatic transmission system with a capacity greater than 10 kW typically includes a cooler. Either an air-cooled or water-cooled heat exchanger can be employed. The heat exchanger should be installed in a low-pressure reservoir return line. All flows, including the pump and motor case drain, must be directed to the heat exchanger inlet.

Case Drains

In pumps and hydraulic motors used in HSTs, fluid leaks internally into their casing. When a pump or motor gets worn or damaged, internal leakage increases. Therefore, a large pump and motor used in an HST typically includes case drain lines connected to the reservoir to relieve internal leakage, prevent leakage fluid from building pressure in the pump and motor casings, and assist in cooling. Moreover, the flow passing through the associated charge relief or flushing valve may be routed to the case drain lines.

Filters

Because piston pumps and motors are commonly used in closed-circuit HSTs, they require high-quality fluid to perform optimally. Filtering is an essential requirement for closed-circuit HSTs.

A strainer with a mesh size greater than 150 microns can be used on the suction side of the charge pump, and a fine filter with a mesh size of 3 to 10 microns can be used on the outlet side. A high-pressure fine filter can also be used in the main transmission loop.

Flushing Valve System

The basic hydrostatic drive is also fitted with an integral or external closed-circuit flushing valve system to direct the hot fluid from the transmission loop to the reservoir, preferably through the series-connected case drain lines of the motor and pump. The following section explains an HST circuit with a flushing valve.

A Closed-circuit HST with a Charge Pump and Flushing Valve System

Figure 3.3(a) shows a typical bidirectional closed-circuit HST with the main bi-directional pump and a motor. It is provided with a charge pump system and a flushing valve system. The charge pump system comprises a charge pump, two check valves (CV1 and CV2), and a charge pressure relief valve (PRV1), as previously explained. The flushing valve system consists of a shuttle valve (pilot-operated directional control valve) and a low-pressure flushing pressure relief valve (PRV2).

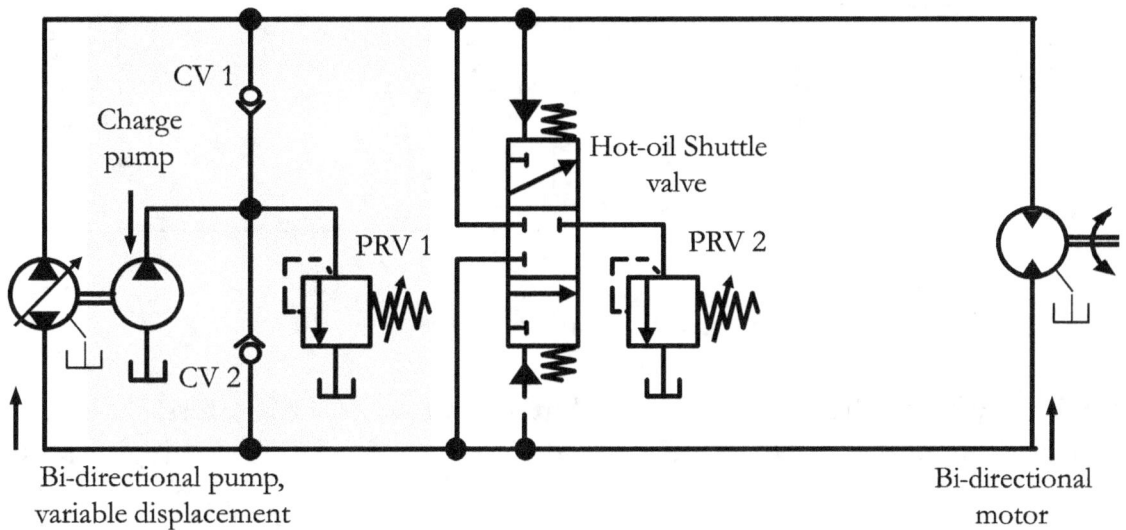

Figure 3.3(a) | A closed-circuit HST

Shuttle Valve

A shuttle valve connects the low-pressure side of the transmission loop to the reservoir through the flushing pressure relief valve PRV2. The shuttle valve is a 3/3-way double-pilot valve. The flushing pressure relief valve (PRV2) is typically set 2 bar below the charge pressure relief valve (PRV1). Two critical positions of the circuit are given in Figures 3.3(b) and 3.3(c).

Position When the Flow is in the Clockwise Direction: Figure 3.3(b) shows the circuit position with the high-pressure side on top, and the shuttle valve is actuated for the left-hand-side envelope. Flow is directed to the reservoir when the low-pressure side pressure exceeds the flushing relief valve (PRV2) setting.

The shuttle valve bleeds about 10% of the flow for cooling. The constant loop replenishment helps prevent heat and contamination from accumulating in the loop. It should also be noted that fluid leakage from the pump and motor can be diverted back to the reservoir, as explained in the next section.

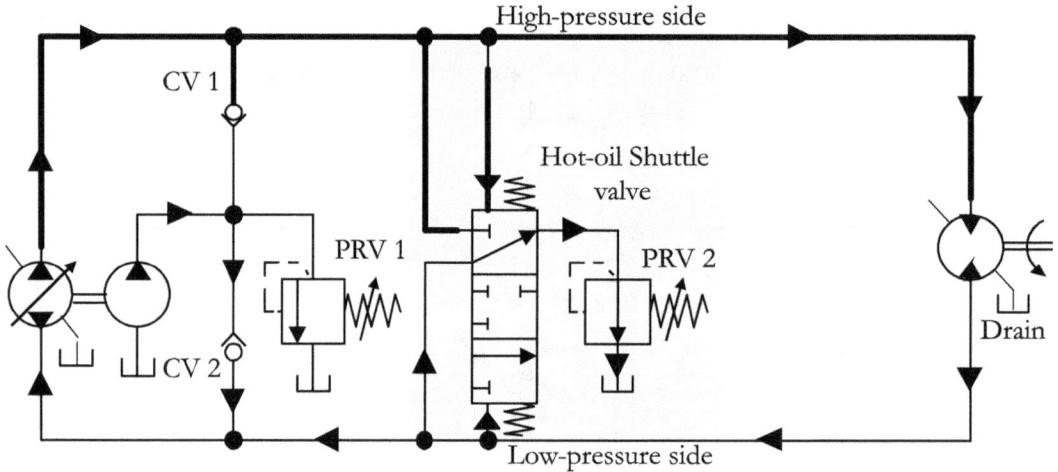

Figure 3.3(b) | A critical position of the closed-circuit HST

Position When the Flow is in Anti-Clockwise Direction: When the swashplate angle is shifted for the reversal of flow around the loop, high-pressure fluid flows through the bottom side of the transmission loop, and hence the shuttle valve is actuated for the right-hand side envelop, as shown in Figure 3.3©.

The flow can also be routed through the case drains in the hydraulic motor and pump before returning to the reservoir for flushing and lubricating the working parts of the motor and pump cases, and increased cooling and filtering, as shown in the Figure. The constant-loop replenishment helps prevent heat and contamination from accumulating within the circuit.

Note that there are many methods for case drain routing.

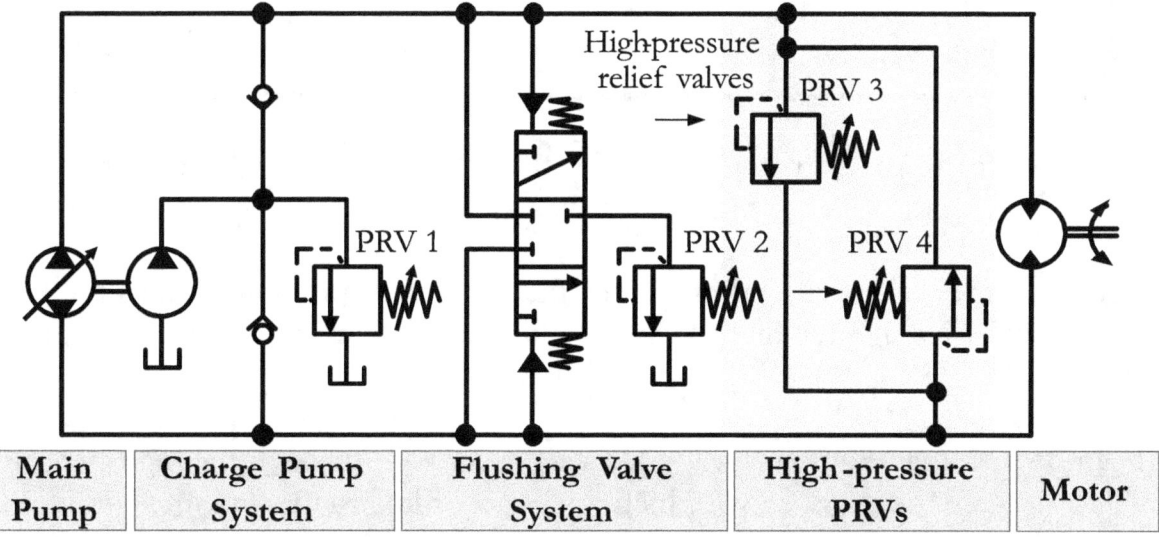

Figure 3.3(c) | A critical position of the closed-circuit HST with flushing flow routed through the case drains of the hydraulic motor and pump

A Closed-circuit HST with a Charge Pump, a Flushing Valve System, and High-pressure Relief Valves

High-pressure pilot-operated relief valves act as protection devices. They limit the system's maximum operating pressure and prevent damage from inadvertent overload on the hydraulic motor. Two pressure relief valves can be used to safeguard the motor during its bi-directional operation.

Figure 3.4(a) | An HST circuit with high-pressure relief valves

A high-pressure relief valve is typically set about 10% above the pump's compensator setting. If the feed pressure rises above the relief valve setting, the fluid bypasses the motor through the high-pressure relief valve.

If the relief valve is connected directly to the reservoir, the transmission loop and the pump's suction side will starve for fluid, promoting cavitation. Therefore, the relief valve must discharge to the opposite side of the loop, for example, through a cross-port connection, as shown in Figure 3.4(a).

An alternative circuit for connecting the high-pressure relief valves is presented in Figure 3.4(b). The pressure relief valves are connected back-to-back and are linked to the check valves in the charge pump circuit, as shown in the figure.

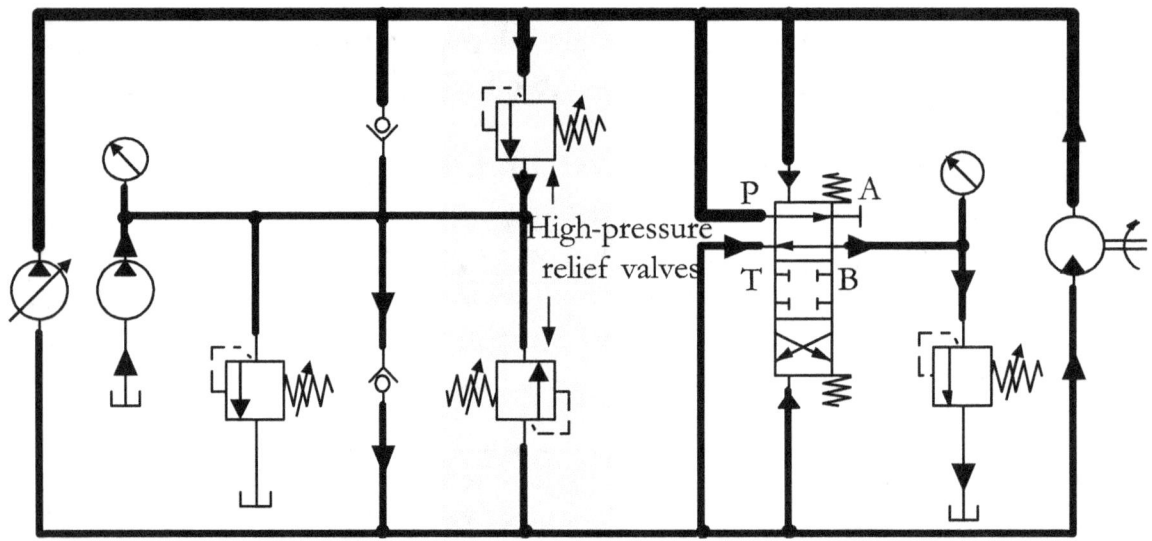

Figure 3.4(b) | An alternative circuit for the connection of high-pressure relief valves

When overpressure occurs, flow moves from the high-pressure side to the low-pressure side through the high-pressure relief valve, then back to the main pump inlet.

It can be seen that the flow always takes an easier path through the high-pressure relief valve and the check valve, which opens according to the pressure conditions in the transmission loop.

A Closed-circuit HST with an Accumulator

In a low-speed, high-torque application employing a closed-circuit hydrostatic transmission system, a high-pressure accumulator can be connected to the high-pressure side of the transmission loop to recover and store energy during load braking or load movement under gravity, and then release the stored energy to generate operating pressure during peak-demand periods. This setup tends to improve system efficiency.

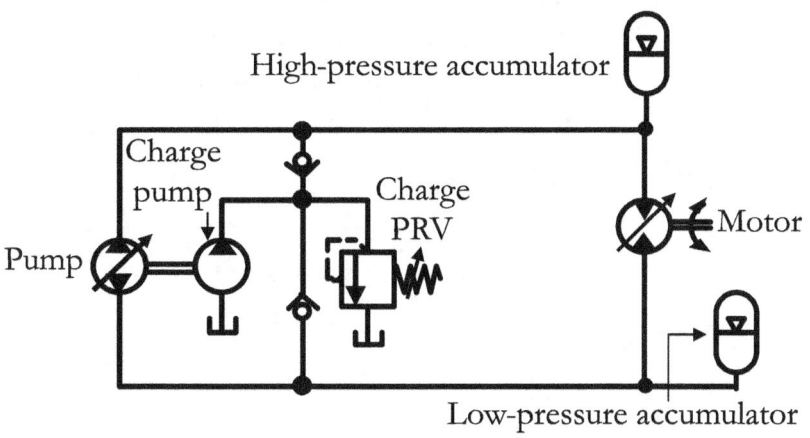

Figure 3.5 | A basic HST with HP and LP accumulators

The same volume of fluid should always be maintained in the transmission loop. Maintaining a constant fluid volume in the loop is not always possible with a high-pressure accumulator, given its charging and discharging phases.

As the fluid volume in the high-pressure accumulator increases, the same amount must be injected into the low-pressure side of the transmission loop, and vice versa. This requirement can be realized by connecting a low-pressure accumulator to the low-pressure side of the transmission loop. A basic closed-circuit HST circuit with high- and low-pressure accumulators is shown in Figure 3.5.

Furthermore, accumulators can be installed on the high- and low-pressure sides of the transmission loop in a closed-circuit HST to dampen pressure fluctuations caused by sudden load or input-supply variations, or to reduce the risk of cavitation on the low-pressure side. Moreover, using an accumulator can reduce the pump size, losses, and cost.

Summary: Component Sizing

Table 3.1 summarises the critical points in closed-circuit hydrostatic systems.

Table 3.1 | Summary points of closed-circuit HSTs

Component	Sizing recommendations
Charge pump flow rate	At least 20% that of the main pump
Total reservoir volume (in litre)	It should be between 0.5 to 1.5 times the maximum flow delivered by the charge pump in lpm (for closed-circuit applications)
Fluid volume in the reservoir, litre	It should be approximately 80% of the total reservoir volume in litre.
Pressure setting, charge PRV	10-35 bar
Pressure setting, Flushing PRV	Pressure setting, charge PRV minus 2 bar
Strainer, charge pump suction	Mesh width may be greater than 150 microns.
Pressure filter, charge pump	Mesh width of 3 to 10 microns
Pressure filter, mainline	A high-pressure fine filter can be used in the main transmission loop.
Heat Exchanger	An HST with a capacity greater than 10 kW is typically equipped with a heat exchanger.
Optimum operating viscosity range	16 to 36 cSt (mm^2/s) 81 to 167 SUS
Viscosity lower to upper limits, typical	10 cSt*to 1000 cSt** [60 SUS to 4635 SUS] *For short periods at a maximum permissible drainage oil temperature of 90°C (194°F) **For short periods upon a cold start
Suction line	-Short and straight -Typically, pressure should not fall below -0.2 bar and should not rise above 2 bar
The direction of rotation of the pump drive	As indicated by a directional arrow
Case drain	Must always drop below the minimum oil level

Chapter 4 | HST Configurations

The elements of an HST system can be configured to maximize the use of space. Accordingly, there are two basic types of configurations used. They are: (1) Close-coupled configuration and (2) Split configuration. The close-coupled configurations can be classified as in-line, U-shaped, or S-shaped.

The pump and hydraulic motor share a common valving surface in the close-coupled configuration, as shown in Figure 4.1(a). All components are enclosed in a common enclosure. This arrangement provides a concise fluid-flow path, thereby eliminating high-pressure fluid leaks.

The close-coupled transmissions are typically used in light-duty applications, where space constraints require compact units. The integrated package can eliminate additional parts and provide economical solutions to application requirements.

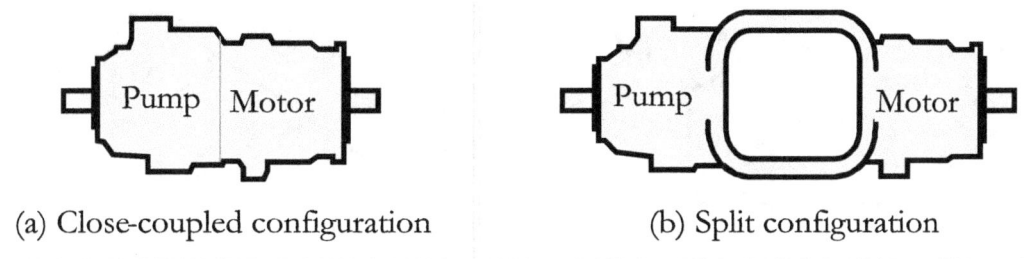

(a) Close-coupled configuration   (b) Split configuration

Figure 4.1 | Schematic diagrams showing two different configurations of hydrostatic systems

In the split configuration, as shown in Figure 4.1(b), a hydraulic pump, reservoir, heat exchanger, filters, valves, and controls are assembled into a power unit, and a remotely mounted hydraulic motor is connected to the power unit via high-pressure hose and steel tubing assemblies.

The split configuration is the most common HST configuration because it enables power to be transmitted to loads in areas that would otherwise be difficult to access. The split configuration is typically used in heavy-duty applications because it is more powerful and provides greater flexibility in system configuration.

Chapter 5 | Types of Hydrostatic Transmission Systems

Hydrostatic transmissions (HSTs) are commonly available with at least three output performance standards. They are: (1) Variable-power, constant-torque transmissions, (2) Variable-power, variable-torque transmissions, and (3) Constant-power, variable-torque transmissions. Typical characteristics of an HST are given in Figure 5.1.

Variable-power, constant-torque transmission

It is based on a variable-displacement pump supplying fluid to a fixed-displacement hydraulic motor under a constant load. The drive speed can be controlled by varying the pump delivery. This type is considered the best general-purpose drive, with the potential for full-speed operation and simple controls.

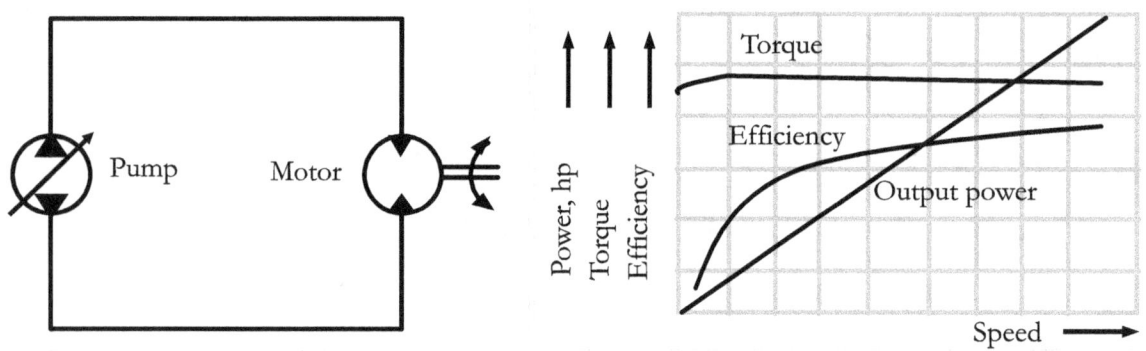

Figure 5.1 | Typical characteristics of an HST

Variable-power, variable-torque transmission

It is based on a variable-displacement hydraulic pump supplying fluid to a variable-displacement hydraulic motor. This transmission type supports variable torque and power. They are adjustable and flexible, but expensive.

Constant-power, variable-torque transmission

It is based on a variable-displacement hydraulic pump with a power limiter, driving a fixed-displacement hydraulic motor. A power controller can regulate the pump's displacement based on the operating pressure, ensuring that the required drive power is not exceeded at a constant drive speed. The advantage of this unit is its efficiency, but its speed range is usually limited.

Chapter 6 | Advantages and Disadvantages of Hydrostatic Transmissions

Hydrostatic transmissions, used in heavy vehicles, compact tractors, and other machinery, offer many advantages. However, these advantages are counteracted by certain disadvantages.

Advantages of HSTs

- A hydrostatic transmission can take all the advantages of the conventional hydraulic system.

- Typically outperforming the mechanical and electrical variable-speed drives, the hydrostatic transmissions offer stepless adjustment of speed, torque, and power.

- An HST offers the ability to operate over a wide range of speeds without changing the speed of the prime mover.

- HSTs provide precise stepless speed control under varying loads, smooth and controllable acceleration/deceleration, and smooth inching operation.

- HSTs can easily control the direction of rotation and stall without damage.

- HSTs have the advantages of fast response and the ability to suppress hydraulic shocks.

- Using proper pumps and hydraulic motors, almost limitless control of any possible movement is possible with the HSTs.

- HSTs offer high efficiency over a wide range of output torques.

- In addition, HSTs also offer the benefits of high power density and flexibility in component arrangement.

Disadvantages of HSTs

- HSTs have lower efficiency (85%) than mechanical transmissions (95%).

- HSTs are costlier than other types of transmissions.

Chapter 7 | Applications of Hydrostatic Transmissions

Hydrostatic transmission (HST) systems are considered excellent for power transmission. HSTs are used in applications that require high power, variable speed, high efficiency, and reliability. HSTs are commonly found on automobiles, marine equipment, and mobile machinery. They can also be used in in-plant industrial systems and wind turbines. Initially, HSTs were used in low-cost systems, such as farm equipment and garden tractors. However, with improved designs, HSTs are suitable for a range of applications, including cranes, excavators, wheel loaders, dozers, tractors, forklifts, and harvesters. A basic classification of HSTs based on the types of applications is presented below:

- Light-duty HST units (< 15 kW) are used on equipment, such as lawn tractors and small machine tools.

- Medium-duty HST units (20 to 40 kW) are used on skid-steer loaders, trenchers, and harvesters.

- Heavy-duty HST units (> 45 kW) are used on agricultural and large construction equipment.

Example 7.1 | A Wind Turbine with an HST

A closed-circuit HST for a gearless wind turbine is shown in Figure 7.1. It usually includes a variable-displacement pump, a variable-displacement motor, a check valve, a pressure relief valve, and an accumulator. Remember, an HST with a variable-displacement pump and motor offers the greatest flexibility.

A charge pump and a heat exchanger can be added to the circuit to compensate for leakage and remove heat. Controllers can also be added to control the speed of the wind turbine and hydraulic motor.

The rotor drives the pump, and the hydraulic motor drives the electric generator. When the rotor rotates, hydraulic fluid in the low-pressure transmission loop is drawn into the pump and delivered to the high-pressure transmission loop. The high-pressure fluid then flows into the motor. Therefore, the motor and the coupled electric generator rotate. As the generator rotates, electricity is generated. The generated electricity is then delivered to the grid.

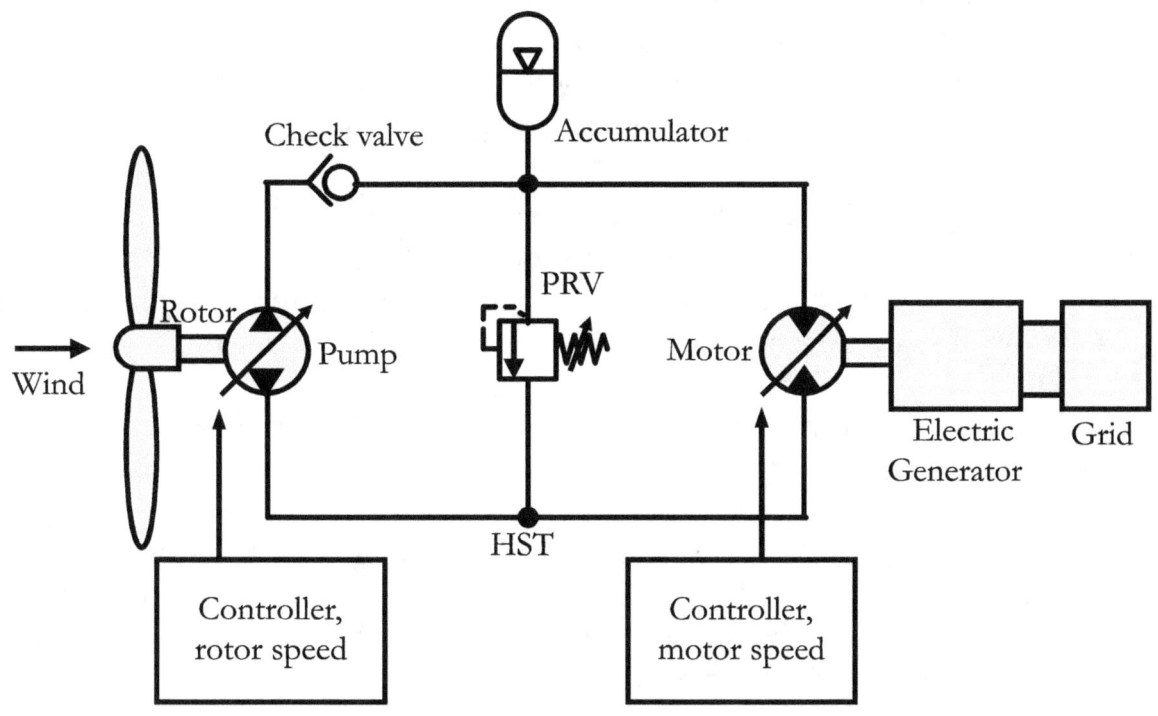

Figure 7.1 | An HST for a wind turbine with an accumulator

The accumulator is connected to the high-pressure side of the loop to store energy when the pressure on that side exceeds the accumulator's pressure. The stored energy in the accumulator can then be released into the system when the loop pressure drops.

The check valve prevents fluid from flowing backward. The pressure relief valve limits the system's maximum working pressure.

Closed-loop controllers regulate the rotor speed and the hydraulic motor speed. It may be remembered that the speed of the hydraulic motor should correspond to the synchronous speed requirement of the electric generator to produce constant-frequency electrical power, regardless of the changes in wind speed or load power.

By adjusting the displacement of either the pump or the motor, the system can achieve a continuously variable gear ratio. This allows the turbine rotor to operate at its optimal tip-speed ratio, defined as the tangential speed of the blade tip relative to the wind speed. This maximises power capture under varying wind conditions, while the generator maintains a constant speed for direct grid connection without complex power electronics.

Example 7.2 | A Vehicle Drive with an HST

Internal combustion (IC) engines power modern vehicles. The engine's power is transmitted through the transmission to the vehicle's axle, driving the wheels. The transmission also adjusts engine speed by engaging and disengaging its gears.

Changing gears disrupts power transmission. This problem can be solved by using a hydrostatic transmission system.

In a hydrostatic transmission, power from the IC engine is transferred to the wheels via a pressurised fluid medium. In this method, as shown in Figure 7.2, the engine drives a pump that delivers pressurised fluid to a hydraulic motor. The motor's rotation drives the coupled wheels.

An HST with a variable-displacement pump and a fixed-displacement motor is widely used as a general-purpose drive, offering a wide range of speeds and ease of control.

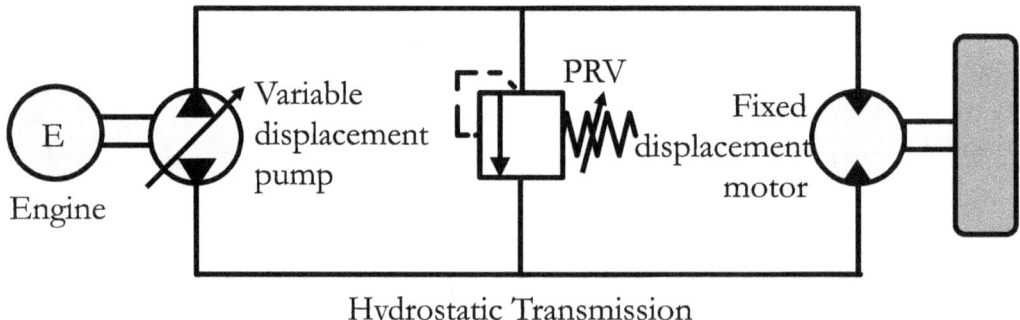

Hydrostatic Transmission

Figure 7.2 | A vehicle with an HST – Schematic diagram

A vehicle drive with a hydrostatic transmission is reliable and easy to manoeuvre. It can operate over a wide range of torque-to-speed ratios and deliver high power per unit of pump/motor displacement. However, the HST system is costly and less efficient than a mechanical transmission.

Chapter 8 | Summary of Relations for Pumps

Positive-displacement pumps come in a wide variety of sizes, flow rates, and power ratings.

Some essential parameters for hydraulic pump operation include pressure rating, priming, slippage, displacement, flow rate, torque, input power, output power, and efficiency.

Figure 8.1 summarizes the important relations of hydraulic pumps in the SI system units.

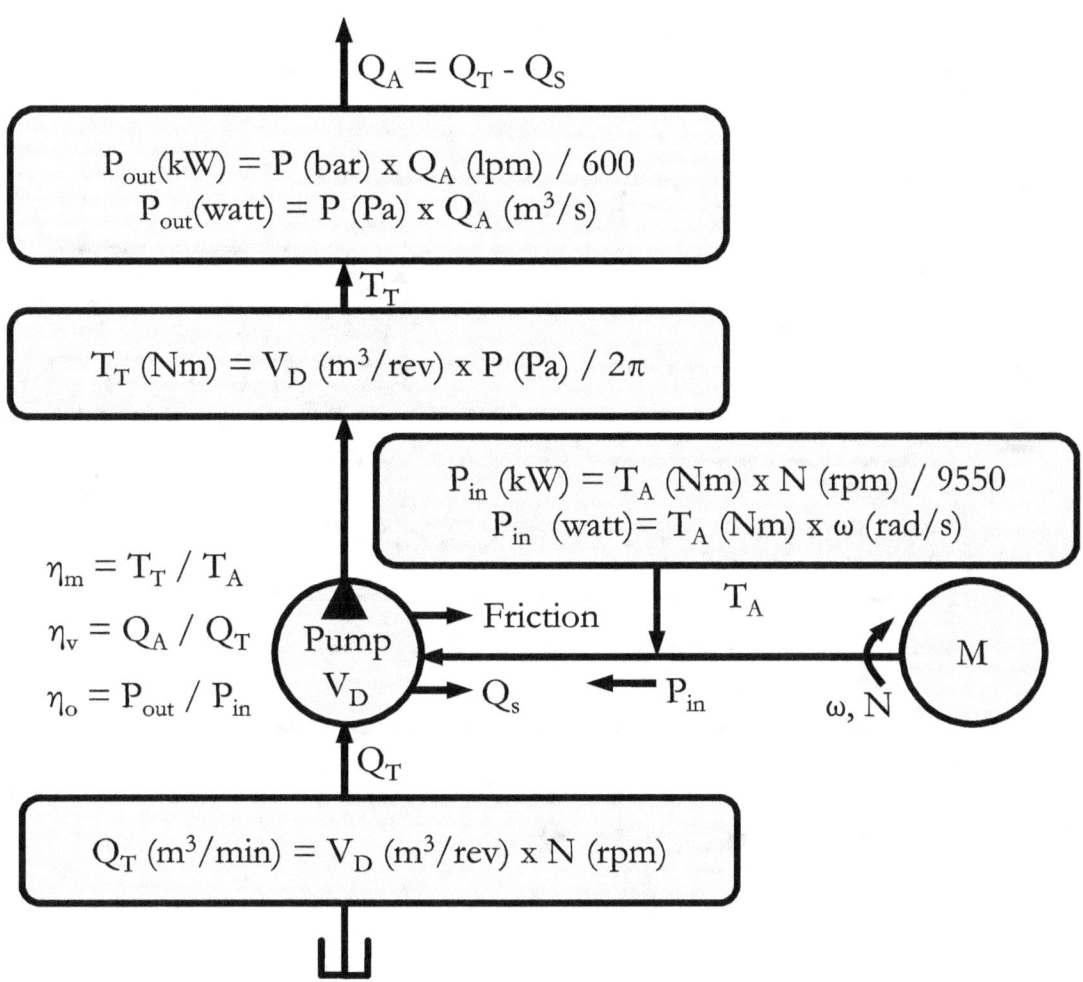

Figure 8.1 | Summary of relations for hydraulic pumps

Chapter 9 | Summary of Relations for Hydraulic Motors

Critical factors for the operation and applications of hydraulic motors include operating pressure, displacement, flow rate, input power, output power, torque output, and efficiency. Figure 9.1 summarizes important relations of hydraulic motors in the SI system units.

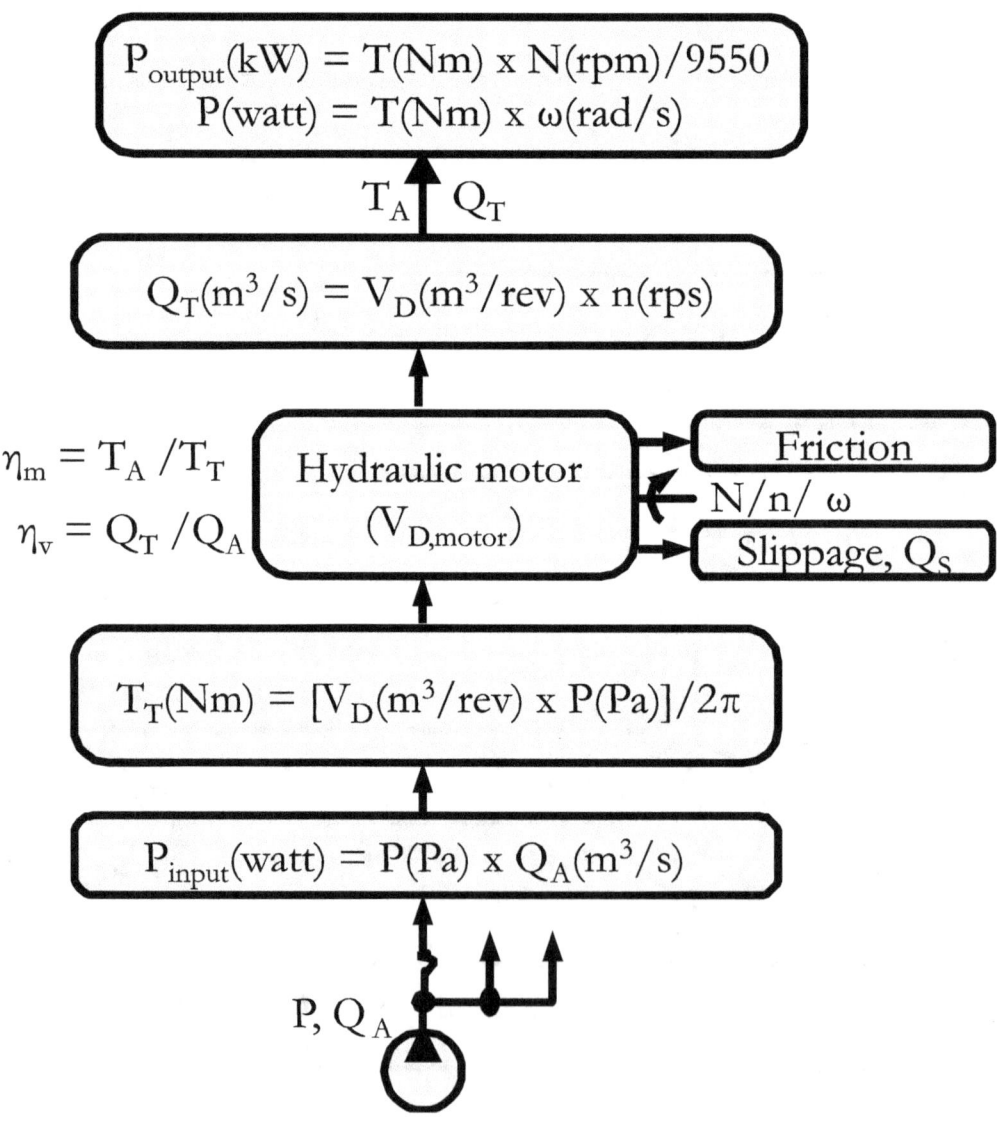

Figure 9.1 | Summary of relations for hydraulic motors

Chapter 10 | A Numerical Problem on an HST

Example 10.1
Hydrostatic transmission (Figure 10.1) with a pump and a hydraulic motor, operating at 70 bar, has the following characteristics for the pump and motor. Determine the displacement of the hydraulic motor and the actual torque developed by the hydraulic motor.

Circuit Diagram

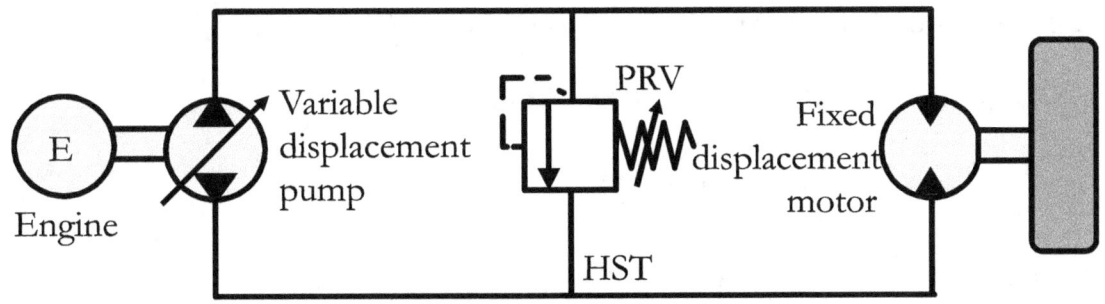

Figure 10.1 | A closed-circuit HST

Pump:
Max. Displacement, $V_{D,P} = 82$ cm^3/rev
Volumetric efficiency, $\eta_{V,P} = 82\%$
Mechanical efficiency, $\eta_{M,P} = 88\%$
Speed, $N_P = 500$ rpm

Motor:
Volumetric efficiency, $\eta_{V,M} = 92\%$
Mechanical efficiency, $\eta_{M,M} = 90\%$
Desired speed, $N_M = 400$ rpm

Displacement, $V_{D,M} =$?

Actual Torque, $T_A =$?

Solution

Theoretical flow rate, pump, $Q_{T,P} = V_{D,P} \times n_P$
$$= 82 \times 10^{-6} \times (500/60)$$
$$= 0.00068 \text{ m}^3/\text{sec}$$

Actual flow rate to the pump, $Q_{A,P} = Q_{T,P} \times \eta_{V,P}$
$$= 0.00068 \times 0.82$$
$$= 0.00056 \text{ m}^3/\text{sec}$$

Actual flow rate of the pump $(Q_{A,P})$ = Actual flow rate to the motor $(Q_{A,M})$

Motor theoretical flow rate, $Q_{T,M} = Q_{A,M} \times \eta_{V,M}$
$$= 0.00056 \times 0.92$$
$$= 0.000515 \text{ m}^3/\text{sec}$$

Motor capacity, $V_{D,M} = Q_{T,M} / n_M$
$$= 0.000515 / (400/60)$$
$$= 0.0000773 \text{ m}^3/\text{rev}$$

Power delivered to the motor, $P_{hyd} = P \times Q_{A,M}$
$$= 70 \times 10^5 \times 0.00056$$
$$= 3920 \text{ Watts} = 3.92 \text{ kW}$$

Mechanical power generated, $P_{Mech} = P_{hyd} \times \eta_{V,M} \times \eta_{M,M}$
$$= 3.92 \times 0.92 \times 0.90$$
$$= 3.246 \text{ kW}$$

Actual Torque developed by motor, $T_A = P_{Mech} / 2\pi n$
$$= 3.246 \times 1000 / (2\pi \, 400/60)$$
$$= 77.49 \text{ Nm}$$

Theoretical torque, $T_T = V_{D,M} \times \Delta P / 2\pi$
$$= 0.0000773 \times 70 \times 10^5 / 2\pi$$
$$= 86.1 \text{ Nm}$$

11 | Objective Type Questions

1. In a typical hydrostatic transmission system:
    a) An electric motor drives a hydraulic pump
    b) An IC engine drives a hydraulic pump
    c) A pump drives a hydraulic motor
    d) A pump drives a hydraulic cylinder

2. Mark the **incorrect** answer
    a) In a hydrostatic transmission system, the speed of the associated hydraulic motor can be controlled infinitely
    b) In a hydrostatic transmission system, the speed reversal of the associated hydraulic motor is quite easy
    c) In a hydrostatic transmission system, the pump is driven at a constant speed
    d) A hydrostatic transmission is a low-cost system

3. Which of the following hydraulic systems draws fluid directly from the reservoir, drives the associated hydraulic motor, and discharges fluid back to the reservoir:
    a) An open-circuit HST
    b) A closed-circuit HST
    c) HST with a charge pump system
    d) An HST with a flushing valve system

4. Which of the hydraulic systems is where the fluid discharged from the hydraulic motor outlet flows directly to the pump inlet:
    a) Open-circuit HST
    b) Closed-circuit HST
    c) Servo valve system
    d) Integrated manifold system

5. In an HST, the hydraulic motor speed can be varied by:
    a) Varying the pressure setting of the charge relief valve
    b) Varying the drive speed of the main pump
    c) Varying the displacement of the main pump
    d) Varying the displacement of the charge pump

6. Mark the **incorrect** statement:
    a) It is not possible to reverse the direction of rotation of the hydraulic motor in a closed-circuit HST
    b) The heat generated in a closed-circuit HST can be dissipated by using the flushing valve system
    c) The torque output of the hydraulic motor in an HST depends on the differential pressure across the motor
    d) Case drain lines in the pump and hydraulic motor in an HST prevent the development of high pressure in the casings of the pump and motor

7. The case pressure in a hydraulic pump used in an HST increases with:
    a) an increase in the wear of pump parts
    b) an increase in the system pressure
    c) increase in leakage
    d) All of the above

8. A charge pump in an HST system:
    a) Compensate for the leakage flows
    b) Controls the speed of the hydraulic motor
    c) Meets the torque requirements of the system
    d) Drives the associated hydraulic motor hydraulically

9. A charge pump system in an HST consists of the following:
    a) A pump and a hydraulic motor
    b) A pump, a shuttle valve, and a PRV
    c) A pump, a reservoir, check valves, a PRV, and a cooler
    d) A pump and high-pressure relief valves

10. A charge relief valve in an HST
   a) Charges the accumulator in the HST
   b) Maintains the pressure in the low-pressure side of the transmission loop
   c) Sets the pressure in the flushing valve system
   d) Maintains the pressure in the high-pressure side of the transmission loop

11. A flushing valve system in an HST:
   a) Uses a shuttle valve and a pressure relief valve
   b) Can direct fluid through the drain lines of the main pump and hydraulic motor
   c) Can direct hot fluid from the transmission loop of the HST to the reservoir through a cooler
   d) All of the above

12. High-pressure relief valves in an HST system:
   a) Absorbs shock pressures
   b) Reduce cavitation in the system
   c) Limits the maximum pressure in the system
   d) Reduce the heat development in the system

13. An accumulator in an HST can:
   a) Dampen pressure fluctuations in the system
   b) Compensate for volume changes in the transmission loop
   c) Reduce cavitation
   d) All of the above

14. An HST system where the hydraulic motor is remotely controlled is most appropriately known as:
   a) Split-configured HST
   b) Close-coupled HST
   c) Open-circuit HST
   d) Closed-circuit HST

Answer key – Objective-type questions:

1-c, 2-d, 3-a, 4-b, 5-c, 6-b, 7-d, 8-a, 9-c, 10-b, 11-d, 12-c, 13-d, 14-a

12 | Review Questions

1. What are the different ways power can be transmitted?
2. Describe the construction, operation, and application of open-circuit hydrostatic transmission.
3. Explain the operation of a closed-circuit hydrostatic transmission (HST) with a neat sketch.
4. Differentiate open-circuit HST and closed-circuit HST
5. Explain the functions of the charge pump in an HST
6. Explain the functions of the low-pressure relief valve in a closed-circuit HST
7. What is the purpose of a heat exchanger in an HST?
8. What is the impact of leakage in an HST?
9. Explain the functions of the flushing valve system in an HST
10. What is the purpose of a shuttle valve in an HST?
11. Explain the function of high-pressure PRVs in an HST
12. Differentiate close-coupled and split HST configurations
13. What are the advantages of hydrostatic transmissions?
14. Give a few applications of hydrostatic transmissions.

13 | Numerical Problems

1. Hydrostatic transmission with a fixed displacement pump and fixed displacement motor operates at a pressure of 138 bar. The motor must drive a load of 350 Nm at 1000 rpm. What is the displacement of the hydraulic motor? Assume the motor's volumetric efficiency is 95%.
{Ans: 0.0001767 m^3/rev}

2. Hydrostatic transmission with a fixed displacement pump and fixed displacement motor operates at a pressure of 100 bar. The pump has a volumetric displacement of 100 cc/rev and runs at 1200 rpm. What are the motor's displacement and output torque if it runs at a speed of 800 rpm? Assume a volumetric efficiency of 88% for the pump. Also, assume the motor's volumetric efficiency is 95%, and its mechanical efficiency is 92%.
{Ans: 0.001254 m^3/rev, 183.58 Nm}

Appendix 1
Specifications of Hydraulic Pumps and Motors in the Metric Units

1. Typical Specifications of Hydraulic Pumps

Table A1.1 | Specifications of pumps

Power (kW)	Displacement (cc/rev)	Max. pressure (bar)	Max. speed (rpm)
69.7	33.3	350	3500
93.3	51.6	350	3100
114.4	69.8	350	2810
134.4	89.0	350	2500
162.7	113.7	350	2350
203.1	165.5	350	2100
325.1	334.4	350	1670

2. Typical Specifications of Hydraulic Motors

Table A1.2 | Specifications of hydraulic motors

Power (kW)	Displacement (cc/rev)	Max. pressure (bar)	Max. speed (rpm)
69.7	33.3	350	3500
93.3	51.6	350	3100
114.4	69.8	350	2810
134.4	89.0	350	2500
162.7	113.7	350	2350
203.1	165.5	350	2100
325.1	334.4	350	1670

Appendix 2

Case Study #1: An Open-circuit Hydraulic Concrete Pump System

Develop a hydraulically driven concrete pump system that uses a drive cylinder and a material cylinder to transfer freshly mixed concrete to a construction site. The material cylinders should alternate between drawing and delivering concrete when activated by the drive cylinders. Using a swing cylinder, a rock valve should connect the material cylinders to the delivery line sequentially to ensure effective concrete discharge.

Solution

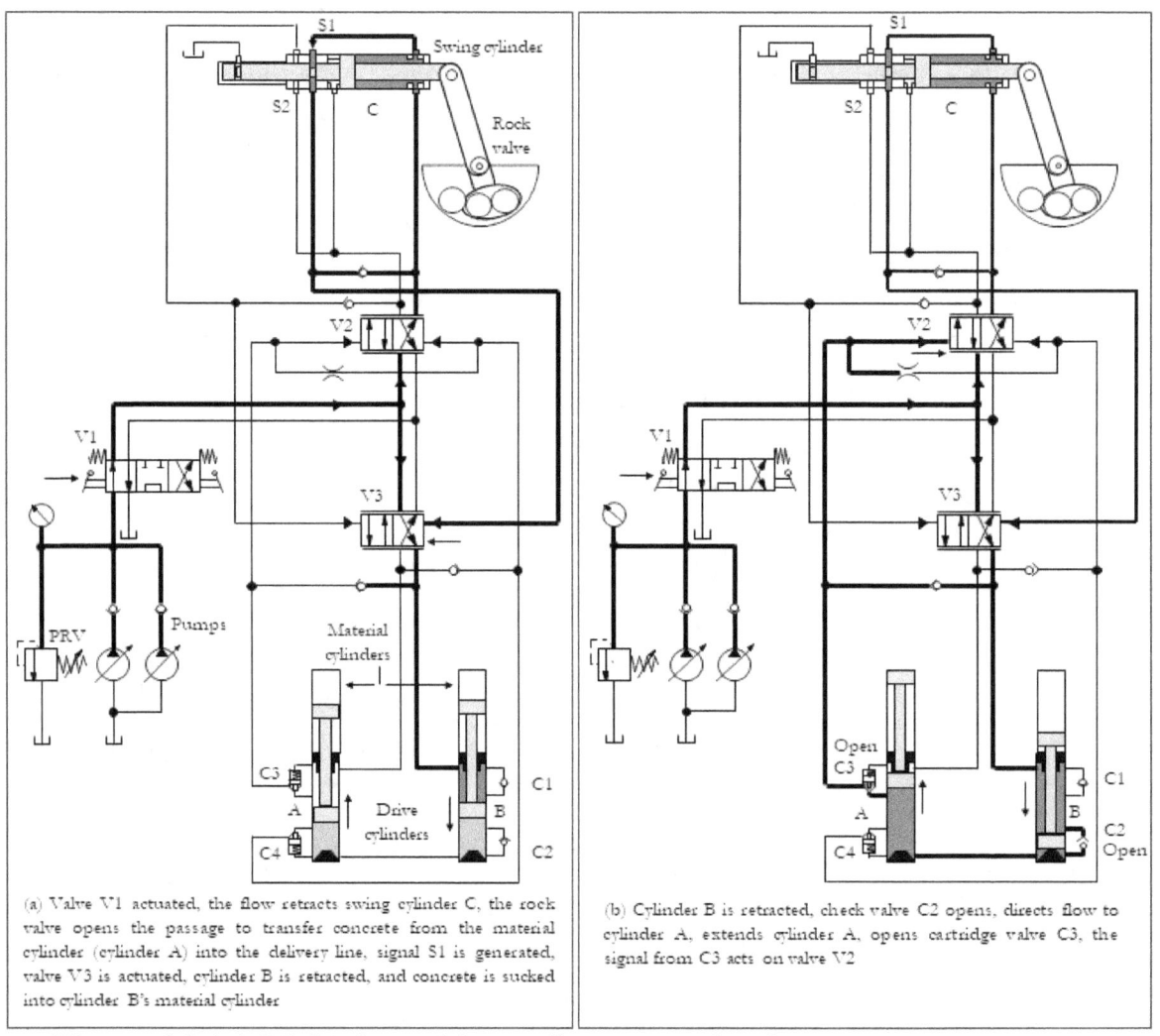

(a) Valve V1 actuated, the flow retracts swing cylinder C, the rock valve opens the passage to transfer concrete from the material cylinder (cylinder A) into the delivery line, signal S1 is generated, valve V3 is actuated, cylinder B is retracted, and concrete is sucked into cylinder B's material cylinder

(b) Cylinder B is retracted, check valve C2 opens, directs flow to cylinder A, extends cylinder A, opens cartridge valve C3, the signal from C3 acts on valve V2

Figure A2.1 | Multiple positions of an open-circuit concrete pump system

The following sections present the various positions of a typical circuit in a concrete pump system. This circuit includes variable-displacement pumps, material cylinders, drive cylinders (A and B), a rock valve, a swing cylinder (C), and a valve system (V1, V2, V3, and PRV).

Figure A2.1(a) shows the circuit's position when valve V1 is activated. Valve V2 directs pump flow to the rod side of swing cylinder C, causing the cylinder to retract and position the rock valve.

This allows the concrete from the material cylinder attached to drive cylinder A to be pushed into the delivery pipeline.

When swing cylinder C retracts fully, it generates signal S1. This signal acts on the right-hand side pilot line of valve V3 and actuates the valve.

As a result, the pump flow is directed towards the drive cylinder B to facilitate its retraction.

When cylinder B retracts, concrete from the hopper is sucked into the material cylinder attached to cylinder B.

Meanwhile, the fluid flow from the piston side of cylinder B is redirected to the piston side of cylinder A.

As a result, cylinder A extends simultaneously, pushing the concrete from its attached material cylinder into the delivery line.

Figure A2.1(b) shows the position of the circuit when cylinder B is fully retracted. At this point, check valve C2 opens, directing the flow towards the piston side of cylinder A through cylinder B.

This creates enough pressure to move cylinder A to its end position and open cartridge valve C3.

The high-pressure signal through the cartridge valve then acts on the left-hand pilot line of valve V2, while fluid from the right-hand pilot line of valve V2 is released to the tank through valve V2 and valve V1.

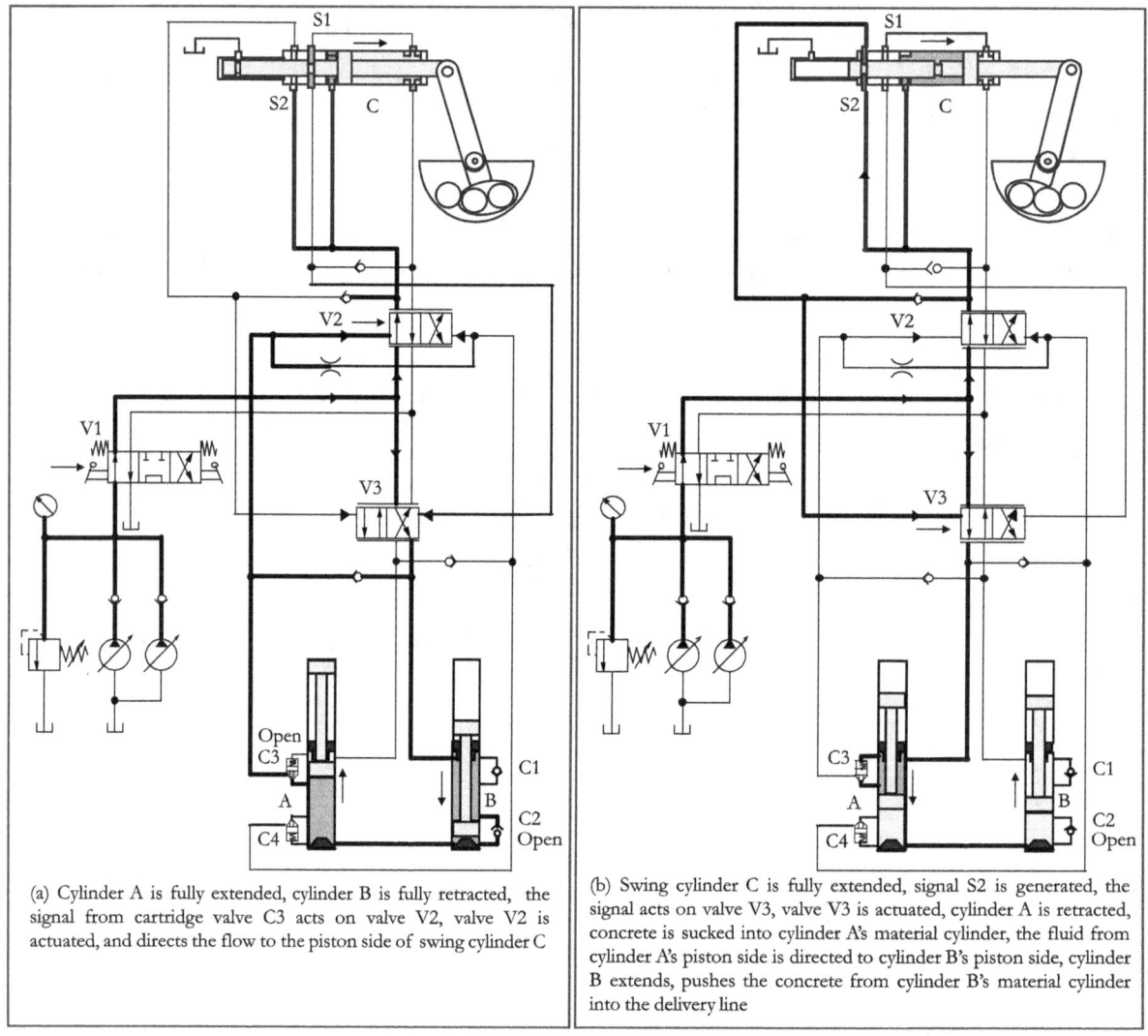

(a) Cylinder A is fully extended, cylinder B is fully retracted, the signal from cartridge valve C3 acts on valve V2, valve V2 is actuated, and directs the flow to the piston side of swing cylinder C

(b) Swing cylinder C is fully extended, signal S2 is generated, the signal acts on valve V3, valve V3 is actuated, cylinder A is retracted, concrete is sucked into cylinder A's material cylinder, the fluid from cylinder A's piston side is directed to cylinder B's piston side, cylinder B extends, pushes the concrete from cylinder B's material cylinder into the delivery line

Figure A2.2 | Multiple positions of an open-circuit concrete pump system

Figure A2.2(a) shows the position of the circuit when cylinder A is fully extended and cylinder B is fully retracted. The signal from cartridge valve C3 acts on the left-hand side pilot line of valve V2. As a result, valve V2 is actuated to its left envelope position. This directs the flow to the piston side of swing cylinder C.

Figure A2.2(b) shows the position of the circuit when swing cylinder C is fully extended. In this cylinder position, signal S2 is generated. The signal acts on the left-hand-side pilot line of valve V3, actuating the valve. As a result, the pump flow is directed toward the drive cylinder A to facilitate its retraction. When cylinder A retracts, concrete from the hopper is sucked into the material cylinder attached to cylinder A. Meanwhile, the fluid flow from the piston side of cylinder A is directed to the piston side of cylinder B. As a result, cylinder B extends simultaneously, pushing the concrete from its attached material cylinder into the delivery line.

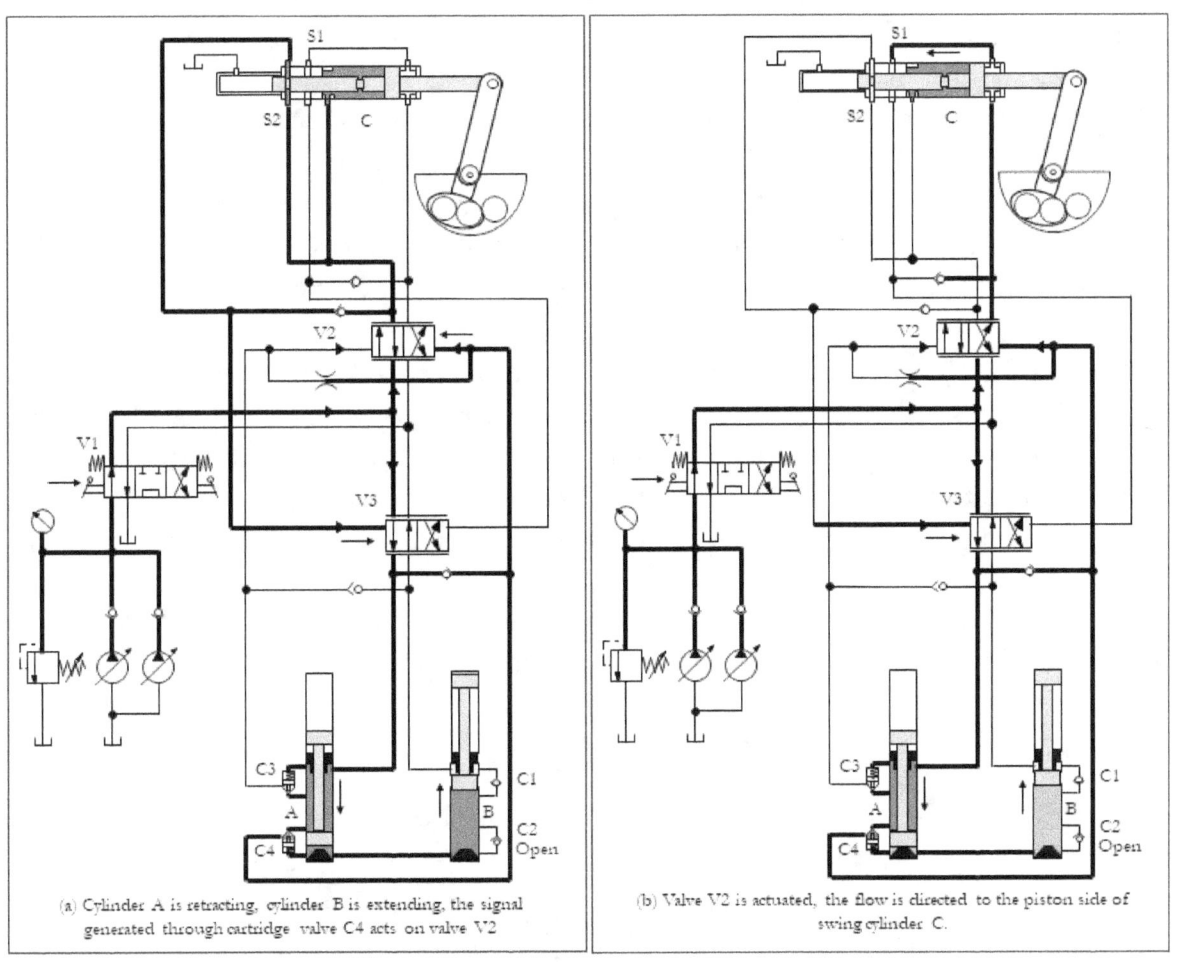

(a) Cylinder A is retracting, cylinder B is extending, the signal generated through cartridge valve C4 acts on valve V2

(b) Valve V2 is actuated, the flow is directed to the piston side of swing cylinder C.

Figure A2.3 | Multiple positions of an open-circuit concrete pump system

The position of the circuit when cylinder A is retracting and cylinder B is extending is given in Figure A2.3(a). Cartridge valve C4, associated with cylinder A, generates a signal that acts on the right-hand side pilot line of valve V2.

The position of the circuit when valve V2 is actuated is shown in Figure A2.3(b). The flow is directed to the piston side of swing cylinder C.

Case Study #2: A Hydrostatic Steering System

Study the operation of a simple hydrostatic steering system for a heavy vehicle.

Explanation

The steering system is responsible for steering a vehicle in the desired direction through gears and linkages. Heavy vehicles generally use hydraulic power steering systems that amplify the torque the operator applies to the steering wheel.

Many off-road vehicles are equipped with hydraulic power steering systems, available in various types and control options. A simple hydraulic power steering system, as shown in Figure A2.4, receives input from the operator via the steering wheel and adjusts the angle of the front wheels via a hydraulic actuator. This makes it easier for the operator to steer the vehicle left or right.

The steering system includes a steering pump, a rotary valve, and a rack-and-pinion gear assembly. The rotary valve, also known as the steering control valve or orbitrol valve, typically has proportional characteristics. A steering cylinder transforms pressurized fluid into mechanical force to adjust the steering position. A double-ended cylinder or two single-ended cylinders can be used for this purpose.

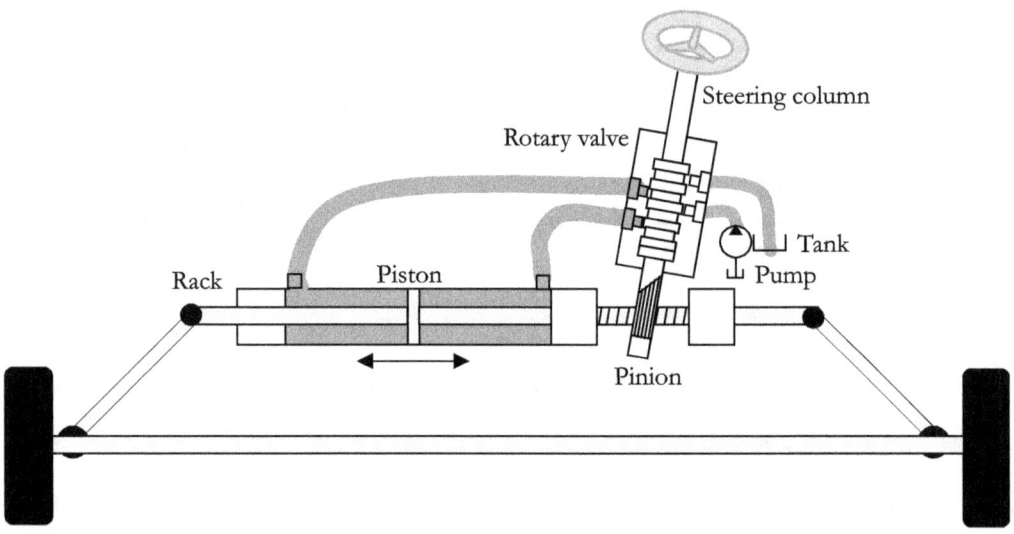

Figure A2.4 | A hydraulic steering system

The pump draws fluid and delivers it to the rotary valve. A rotary valve is a highly sensitive metal casing with strategically placed holes that control the direction of the fluid flow. It diverts fluid from the pump into the steering cylinder to maintain the desired steering position. A properly sized valve meters fluid volume in response to steering wheel rotation.

When the steering wheel is turned in one direction, the rotary valve opens accordingly, allowing the fluid to flow into one side of the piston. This turns the front wheels, changing the vehicle's direction. When the steering wheel is turned in the opposite direction, the rotary valve opens in the opposite way, allowing the fluid to flow into the opposite side of the piston. This also turns the front wheels, changing the vehicle's direction.

Note that hydraulic flow occurs through the valve only when the steering wheel moves. When the movement stops, a feedback mechanism blocks the flow to the piston. In the normal position of the rotary valve, fluid flows to the tank.

A Typical Circuit for a Hydrostatic Power Steering System

Here is an overview of a common hydraulic steering circuit, illustrated in Figure A2.5. The circuit features a variable-displacement pump that controls a steering cylinder via a rotary sliding valve. The system also includes an inlet check valve, anti-cavitation check valves, and cylinder port relief valves.

The anti-cavitation check valves draw fluid from the tank when a vacuum develops in the cylinder's fluid chambers, preventing cavitation and protecting the steering circuit.

Meanwhile, the cylinder port relief valves safeguard hose lines from pressure surges caused by ground forces on the steering axle.

Lastly, the inlet check valve prevents backflow through the steering unit when the pressure on the cylinder side exceeds that on the inlet side.

The closed-center rotary slide valve has a spool that moves within a sleeve, enabling proportional flow metering. A proportional solenoid operates the valve, causing the spool to rotate and enabling flow proportional to the current produced by the steering wheel's rotation. The valve also features a feedback mechanism with a rotary motor unit that controls the valve spool to match the valve sleeve's movement through pivoted mechanical connections.

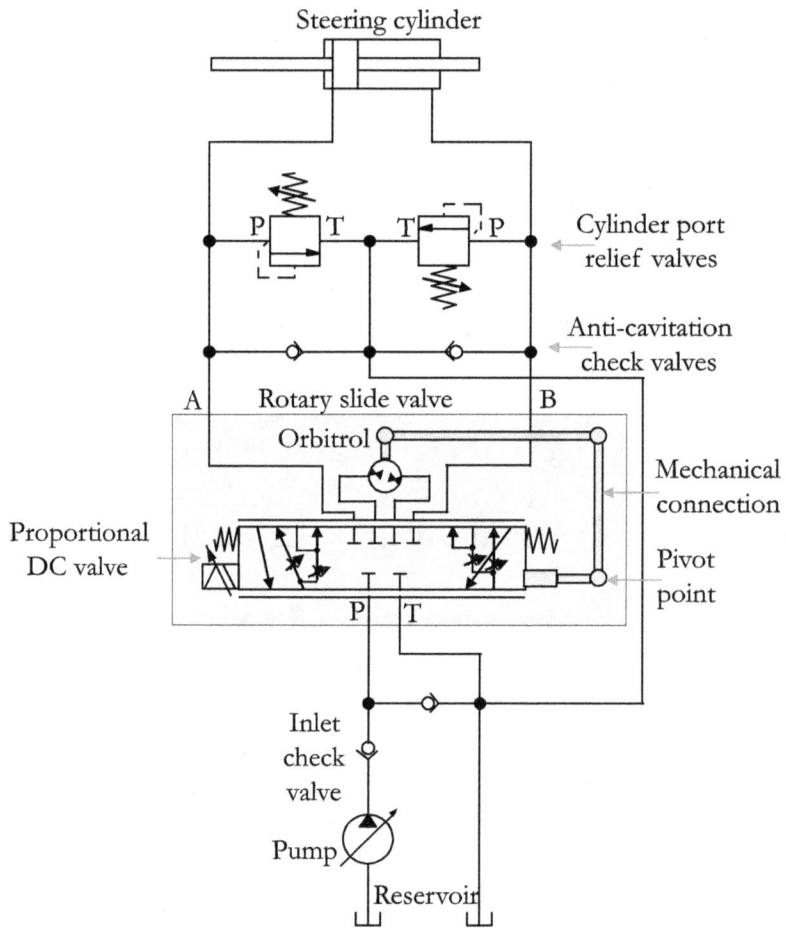

Figure A2.5 | A typical circuit of a power steering system

When the operator slightly turns the steering wheel to the right or left, the servo valve opens just enough to direct the fluid to the cylinder, which turns the wheels in the desired direction.

As the valve opens, it also diverts some of the fluid to the orbitrol motor and triggers the mechanical connections to push the sleeve to match the spool's movement, thereby closing the valve. This ensures that the valve output remains proportional to the difference between the operator-provided command signal and the motor's actual output.

Case Study #3: Typical Displacement Control of Bidirectional Variable-displacement Axial-piston Pump

Explanation
A typical hydrostatic drive with a bidirectional variable-displacement axial piston pump is given in Figure A2.6. The axial-piston pump can swivel to both sides. An electronic displacement control method adjusts the pump displacement in both flow directions.

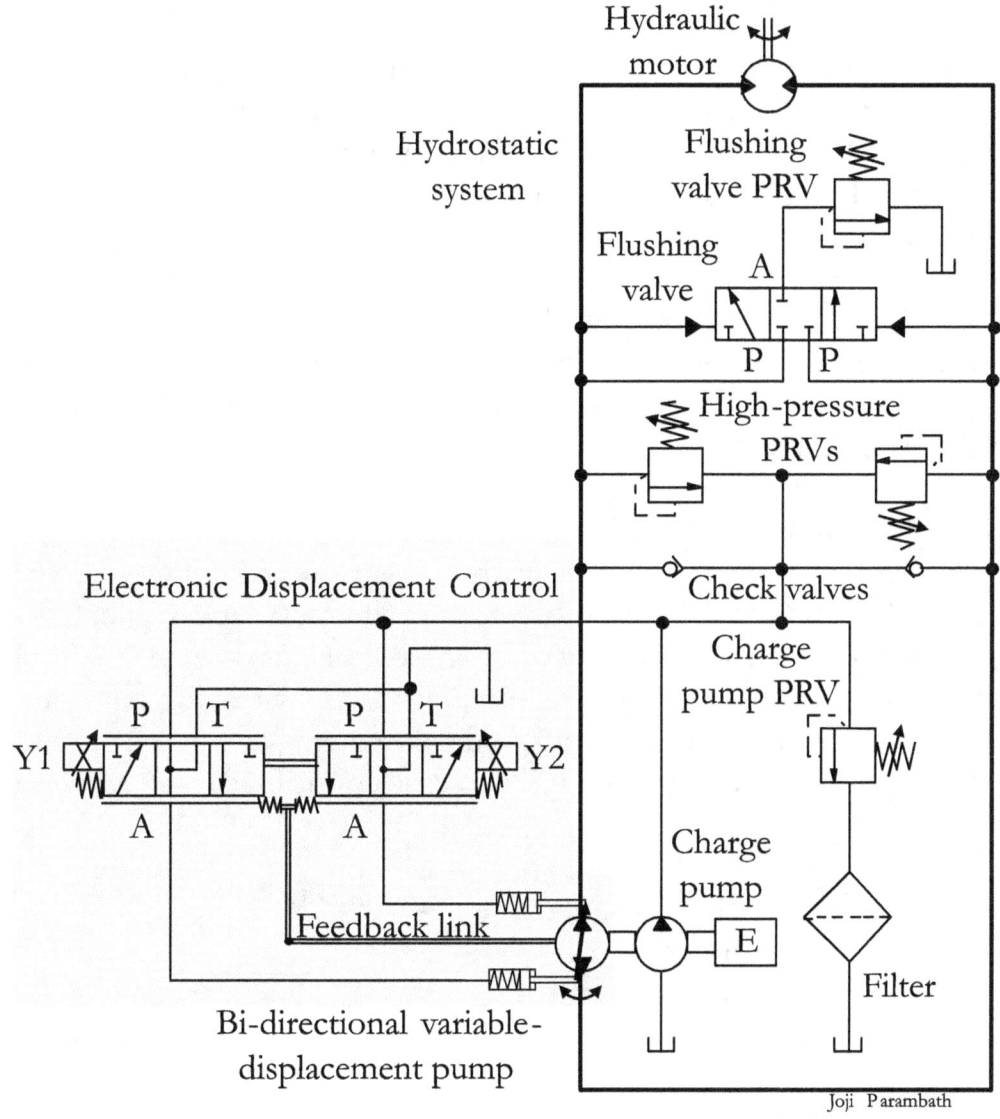

Figure A2.6 | A typical displacement control of a bi-directional pump

In the closed-circuit hydrostatic system, a bidirectional variable-displacement axial-piston pump hydraulically drives a fixed-displacement piston motor. The charge pressure circuit, with a low-pressure charge pump, a small fluid-filled reservoir, a relief valve, and two check valves, compensates for the leakage flows from the pump and the motor's case drain ports. The flushing valve system, with a shuttle valve and a low-pressure flushing pressure relief valve, directs hot fluid from the transmission loop to the reservoir, preferably through the series-connected case drain lines of the motor and pump. High-pressure, pilot-operated relief valves limit the system's maximum operating pressure.

The pump displacement can be regulated using an electronic displacement control system. The displacement control consists of a proportional valve system controlled by proportional solenoids Y1 and Y2. The sliding spools of the proportional valves are mechanically linked and move within sleeves. The valve also features a feedback mechanism that controls the spool to track the sleeve's motion via mechanical connections. The swash plate angle and pump displacement in either pump-flow direction can be controlled by the current signals from the proportional solenoids in response to operator commands. The swash plate's motion activates the mechanical connections, which drive the sleeve to track the spool's motion.

This Topic Requires Further Study
Based on their displacement characteristics in response to the current signal, displacement controls can also be of the following types: forward-neutral-reverse (FNR) electric controls, automotive controls, etc. For example, the three-position FNR electric control uses an electric input signal to switch the pump to a full-stroke position. Typical characteristics of displacement Vs current or voltage for electronic displacement control and FNR electric control are given in Figure A2.7.

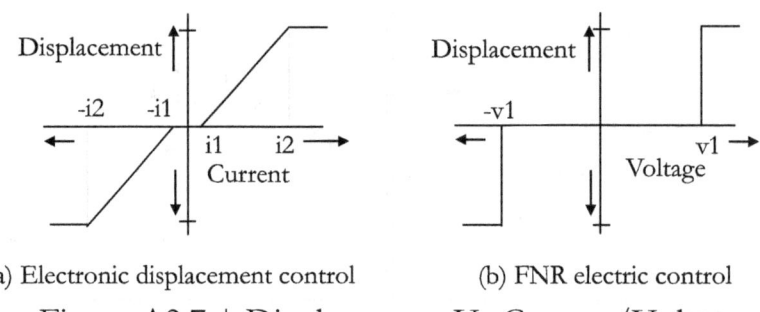

(a) Electronic displacement control   (b) FNR electric control

Figure A2.7 | Displacement Vs Current/Voltage

Appendix 3

Important Parameters of Hydrostatic Transmissions

Parameters	Typical values
Pump displacement	6, 10, 21 cc
Pump swash plate angle	15 degrees
Input speed, rated	3000, 3600 rpm
Input power, maximum	1.1, 2.2, 3.7, 7.4, 11.0 kW
Motor displacement	6, 10, 21 cc
Motor swash plate angle	15 degree
Output speed	3000, 3600 rpm
Rotation, motor	Bidirectional/Clockwise / Counterclockwise
Maximum output torque	9.8, 23.4, 49.2, 72.1 Nm
Charge pump displacement	1.9, 2.1, 3.0 cc
Charge pressure	3 to 5 bar
Inlet pressure, charge pump	0.8 bar(a)
Reservoir	Integrated / External
Reservoir, size	2, 5 litre
Fluid volume in the reservoir	250, 450, 550, 700 cc
Filtration	Integrated / External
Pipe/Tube/Hose size	> 3/8 in
Mounting	As required
Ports	As required
High-pressure relief valve	
System pressure, rated	105, 150, 210 bar
System pressure, maximum	150, 175, 210, 245 bar
Case pressure, rated	0.3 bar
Fluid viscosity	12 – 60 cSt
Fluid temperature	80°C

Appendix 4

Different methods can be used to vary the displacement of an axial piston pump. The following sections explain several displacement-control methods.

Components Used in Variable-displacement Pump Systems

The symbols for components used in variable-displacement pump systems are shown in Figure A4.1.

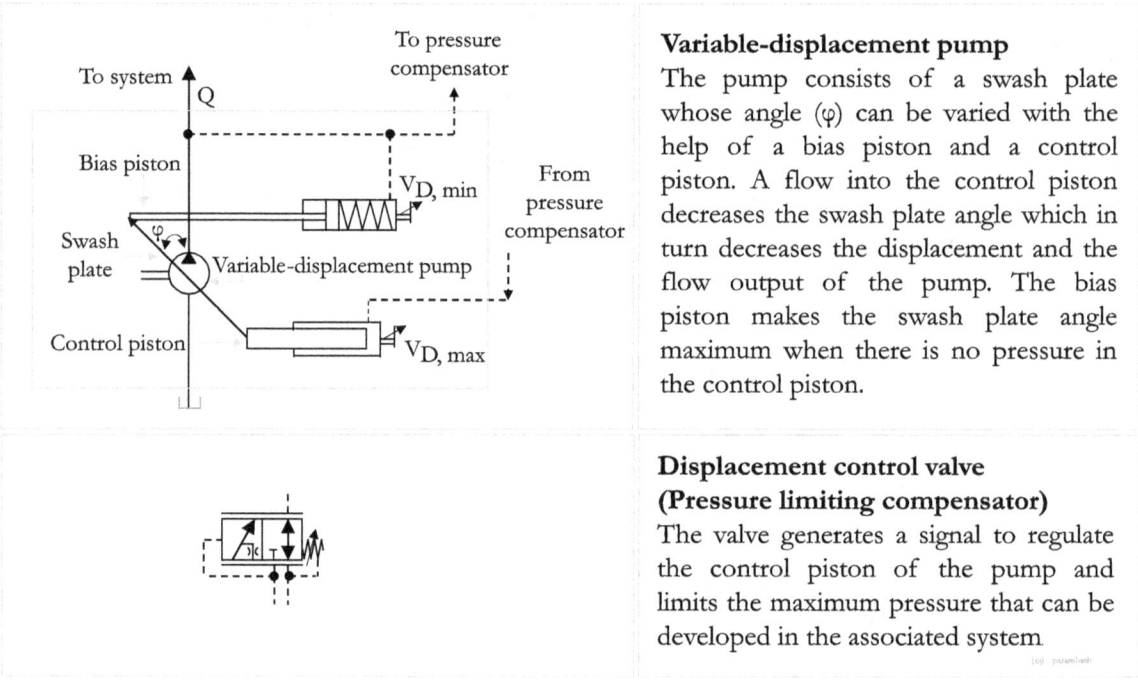

Variable-displacement pump
The pump consists of a swash plate whose angle (φ) can be varied with the help of a bias piston and a control piston. A flow into the control piston decreases the swash plate angle which in turn decreases the displacement and the flow output of the pump. The bias piston makes the swash plate angle maximum when there is no pressure in the control piston.

Displacement control valve (Pressure limiting compensator)
The valve generates a signal to regulate the control piston of the pump and limits the maximum pressure that can be developed in the associated system.

Figure A4.1 | Components used in variable displacement pump systems

Displacement (Maximum): The pump's maximum volumetric displacement can be adjusted using a mechanical stop screw called $V_{D,\,max}$. Turning the screw inward decreases the pump displacement, while turning it outward increases it.

Displacement (Minimum): The pump's minimum volumetric displacement can be adjusted using a mechanical stop screw called $V_{D,\,min}$. Turning the screw inward increases the pump displacement, while turning it outward decreases it.

Variable displacement Pumps

A variable-displacement pump consists of a set of cylinders and pistons, a swash plate, a bias piston, and a control piston. The cylinder block rotates when driven by the prime mover.

The pistons move in and out of their cylinder bores as they slide over the angled swash plate.

When the piston moves out, the fluid is drawn into the cylinder bores from the associated reservoir. The fluid is trapped and then moved from the pump's suction port to its delivery port. The flow is then delivered out.

The bias piston tends to keep the swash plate angle at its maximum displacement.

The control piston tends to keep the swash plate angle at a lower displacement position, as required by the system.

The displacement and, hence, the output flow of an axial-piston hydrostatic pump can be infinitely varied by controlling the swash plate angle of the pump in response to an operator command.

Pump Displacement Control Methods

The displacement angle of a variable-displacement pump can be controlled mechanically by a lever attached to the swash plate, or servo-controlled via a displacement control valve (pressure-limiting compensator).

Direct displacement method: In this method, an operator applies a force to the lever, which moves the swash plate, possibly via a linkage assembly. The force required to operate the lever depends on the system pressure acting on the swash plate. Therefore, this method is generally used only for light- to medium-duty pumps.

Servo-controlled Method: The method uses a control piston to move the swash plate under pressure. The inputs for servo control can be hydraulic or electrical signals. A circuit diagram of a servo-controlled variable-displacement pump system with a displacement-control valve (pressure-limiting compensator) is shown in Figure A4.2.

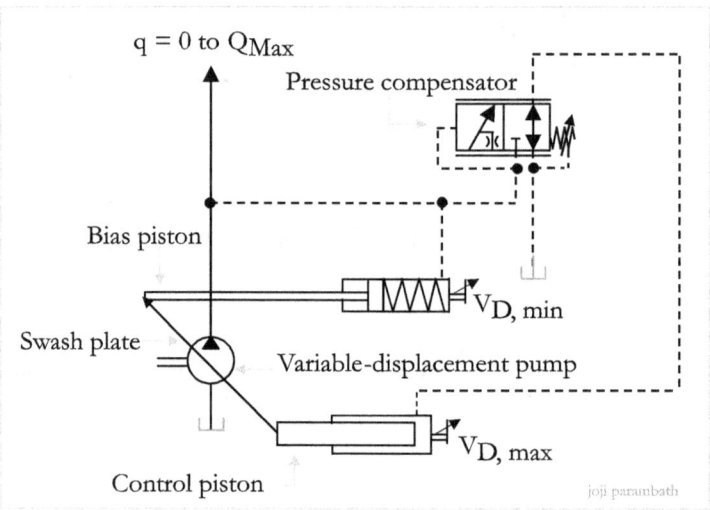

q = 0 to Q_{Max}

Pressure compensator

Bias piston

$V_{D, min}$

Swash plate

Variable-displacement pump

Control piston

$V_{D, max}$

joji parambath

Figure A4.2 | A basic circuit of a servo-controlled variable displacement pump

Remember, the system's life decreases substantially at higher pressures. Therefore, the circuit requires a means to limit or control the maximum pressure in the system.

A pressure relief valve (PRV) can limit the pressure. However, a PRV consumes excess power and generates excessive heat.

A pressure compensator can be used in the system to limit pressure and mitigate power loss and heat generation.

With a proportional characteristic, a displacement control valve controls the pump output flow and limits the pressure to a preset value. The valve constantly senses the pressure at the pump outlet. When the pump outlet pressure increases, the valve throttles the flow and, in proportion, applies a pressure signal to the control piston to reduce pump displacement and, hence, flow output. That is, the valve alters the displacement of the associated variable-displacement axial piston pump under high system pressure.

When the pump outlet pressure reaches the preset level, the pump displacement and flow are reduced to the minimum required to maintain that pressure. The pump produces only enough flow to compensate for internal leakage.

14 | References

1. 'Industrial Hydraulic Systems – Theory and Practice', by Joji Parambath, Universal Publishers, Boca Raton, USA

2. Article on 'Engineering Essentials: Hydrostatic Transmissions' by Alan Hitchcox, https://www.hydraulicspneumatics.com/fluid-power-basics/hydraulics/article/21882804/engineering-essentials-hydrostatic-transmissions

3. Article on 'HYDROSTATIC TRANSMISSIONS', Tandem Hydraulics Pvt Ltd

4. Article on 'Modeling and Control of a 600 kW Closed Hydraulic Wind Turbine with an Energy Storage System' by Liejiang Wei, Zengguang Liu, Yuyang Zhao, Gang Wang, and Yanhua Tao

5. BD Series Hydrostatic Transmission – Technical Information, Sauer-Danfoss, www.sauer-danfoss.com

6. Catalog Model TECHNICAL INFORMATION MANUAL FOR THE A10V0 VARIABLE DISPLACEMENT PUMPS, 96-MOD3A-70196, COMPU-SPREAD, Basic Technologies

7. Danfoss: Application Manual – Transmission Circuit Recommendations, powersolutions.danfoss.com

8. Document on 'Controllers devices DR, DP, FR, and DFR for Variable axial piston pump (A)A4VSO', RE-A 92060 Edition: 11.2017, Bosch Rexroth Corporation, Mobile Applications, Fountain Inn, SC USA

9. Document on 'Hydraulic - Training Axial Piston Units, Basic Principles', RE 90600/01.98, Mannesmann Rexroth

10. Document on 'HYDROSTATIC TRANSMISSIONS', Number 75, TECHNI/TIPS, Lubrication Engineers Technical Department, 300 Bailey Ave, Fort Worth, TX 76107, http://www.le-inc.com

11. Document on 'Hydrostatic Transmissions', Penton Media, Inc.

12. Eaton Hydraulics Training Services, Mobile Hydraulic Manual, 2010

13. Paper on 'DESIGN OF A POWER REGENERATIVE HYDROSTATIC WIND TURBINE TEST PLATFORM', by Biswaranjan Mohanty, Feng Wang, and Kim A Stelson, Center for Compact and Efficient Fluid Power, Department of Mechanical Engineering, University of Minnesota, Minneapolis, MN 55455, US

14. Publications Department of Womac Machine Supply Company, Industrial Fluid Power, Volume 3, Third Edition

Fluid Power Educational Series Books

1. Pneumatic Systems and Circuits -Basic Level (In the SI Units)
2. Industrial Pneumatics -Basic Level (In the English Units)
3. Pneumatic Systems and Circuits -Advanced Level
4. Electro-Pneumatics and Automation
5. Design of Pneumatic Systems (In the SI Units)
6. Design Concepts in Pneumatic Systems (In the English Units)
7. Maintenance, Troubleshooting, and Safety in Pneumatic Systems
8. Industrial Hydraulic Systems and Circuits -Basic Level (In the SI Units)
9. Industrial Hydraulics -Basic Level (In the English Units)
10. Hydraulic Fluids
11. Hydraulic Filters: Construction, Installation Locations, and Specifications
12. Hydraulic Power Packs (In the SI Units)
13. Power Packs in Hydraulic Systems (In the English Units)
14. Hydraulic Cylinders (In the SI Units)
15. Hydraulic Linear Actuators (In the English Units)
16. Hydraulic Motors (In the SI Units)
17. Hydraulic Rotary Actuators (In the English Units)
18. Hydraulic Accumulators and Circuits (In the SI Units)
19. Accumulators in Hydraulic Systems (In the English Units)
20. Hydraulic Pipes, Tubes, and Hoses (In the SI Units)
21. Pipes, Tubes, and Hoses in Hydraulic Systems (In the English Units)
22. Design of Industrial Hydraulic Systems (In the SI Units)
23. Design Concepts in Industrial Hydraulic Systems (In the English Units)
24. Maintenance, Troubleshooting, and Safety in Hydraulic Systems
25. Hydrostatic Transmissions (HSTs) (In the SI Units)
26. Concepts of Hydrostatic Transmissions (In the English Units)
27. Load Sensing Hydraulic Systems (In the SI Units)
28. Concepts of Load Sensing Hydraulic Systems (In the English Units)
29. Electro-hydraulic Proportional Valves
30. Electro-hydraulic Servo Valves
31. Cartridge Valves
32. Electro-hydraulic Systems and Relay Circuits
33. Practical Book: Pneumatics - Basic Level
34. Practical Book: Electro-pneumatics - Basic Level
35. Practical Book: Industrial Hydraulics – Basic Level
36. Programmable Logic Controllers and Programming Concepts
37. Compressed Air Dryers
38. Hydraulic Circuits – Identification of Components and Analysis

For more details, please visit **https://jojibooks.com.**

About the Author

Joji Parambath has been a resource person in Pneumatics, Hydraulics, and PLC for over 25 years. He has trained professionals from various industries, faculty members, and engineering students throughout his career.

Joji is the main trainer at Fluidsys Training Centre in Bangalore, India. This centre offers training in Pneumatics and Hydraulics. Joji has authored a series of 38 books that aim to update and restructure the existing ones.

Joji would like to express his gratitude to all the trainees who actively engaged with him and offered valuable suggestions during the training programs, which motivated him to create this series of books. If you have any feedback, please direct it to joji.p@hotmail.com.

25th August 2023

www.ingramcontent.com/pod-product-compliance
Lightning Source LLC
Chambersburg PA
CBHW082152200526

45794CB00008B/3267

* 9 7 9 8 8 5 9 1 0 4 0 7 9 *